Clustering for Big Data

Towards Generalized VAT-style Class-independent Unsupervised Fuzzy Clustering for Big Data

ARASH ABADPOUR

Arash Abadpour (arash@abadpour.com) is with Fio Corporation.

The author acknowledges that this text has been submitted for publication and that this file will be deleted upon publication.

Contents

1	**Introduction**	**1**
2	**Literature Review**	**3**
3	**Method**	**13**
	3.1 Modeling Framework	13
	3.2 Model Preliminaries	14
	3.3 Assessment of Loss	15
	3.4 Single-Cluster Clustering	18
	3.5 Cluster Space Sampling	19
	3.6 Cluster Aggregation	21
	3.7 Determination of U and λ	22
4	**Experimental Results**	**25**
5	**Conclusions**	**43**

List of Figures

3.1 The process of determining λ and U. (a) Selection of φ_f. (b) Selection of φ_p. 23

4.1 Geometry of the search space for an arbitrary 2de problem instance. 27

4.2 Cluster dissimilarity matrixes corresponding the input set of data items utilized in Figure 4.1. (a) DM. (b) RDM. 27

4.3 visualizations of the MST corresponding to the input set of data items utilized in Figure 4.1. (a) Successive MST edge lengths. (b) Prominence/Weight graph. 28

4.4 Set of clusters produced by the developed method for the data shown in Figure 4.1. 30

4.5 Convergence paths of the cluster seeds corresponding to the clusters shown in Figure 4.4. (a) Two dimensional view from top. (b) Three dimensional view of cluster representations and the associated cost function values. 30

4.6 Membership levels of the reordered input data items to the converged clusters shown in Figure 4.4. 31

4.7 Assessment of the cluster dissimilarity measure utilized in this work for an arbitrary 2de problem instance. (a) and (b) Two cluster candidates produced by the developed algorithm. (c) Reordered membership graphs for the two clusters carried in (a) and (b)... 33

4.8 Repeatability assessment for the 2dl problem instance shown in (a). (b), (c), (d), (e) and (f) exhibit the results of five independent executions of the developed algorithm. 34

4.9 Impact of $\hat{C}$ on the outputs of the developed algorithm for a 3dpp problem instance. Top: Converged clusters. Bottom: Classification results. (a) $\hat{C} = 5$. (b) $\hat{C} = 10$. (c) $\hat{C} = 20$. (d) $\hat{C} = 40$. (e) $\hat{C} = 80$. (f) $\hat{C} = 100$. Input data courtesy of *Epson Edge, Epson Canada Limited.* 36

4.10 Number of discovered clusters and the elapsed time for the experiments shown in Figure 4.9. (a) Number of clusters. (b) Elapsed time. 37

4.11 Sample results generated by the proposed algorithm for problem instances corresponding to the six problem classes carried in Table 4.1. (a) 2dc. (b) 2de. 38

4.12 Sample results generated by the proposed algorithm for problem instances corresponding to the six problem classes carried in Table 4.1. (a) 2dl. (b) 3dpp. Input data courtesy of *Epson Edge, Epson Canada Limited*. 39

4.13 Sample results generated by the proposed algorithm for problem instances corresponding to the six problem classes carried in Table 4.1. (a) ics. (b) ighe. 40

4.14 Undesired results generated by the proposed algorithm. (a) 2dc. (b) 2de. 41

Abstract

It is a prerequisite to many applications in different fields to separate a set of data items into homogenous clusters. In this context, a data item may be a member of $\mathbb{R}^k$ or a complex mathematical entity encompassing several properties. Homogeneity, too, is a general concept which is defined differently in different contexts. In effect, the ability to work with abstract models for data items and clusters has important economical benefits, in terms of not only the reusability of the algorithms, but also the reuse of actual computer code. Visual Assessment of Cluster Tendency (VAT) is a discovery mechanism which has been shown to work desirably within the context of prototype-based clustering. However, VAT has been shown to suffer from high costs of operation, especially in the context of Big Data. While the research community has invested heavily on proposing alternatives to VAT, a generic cost-effective unsupervised cluster discovery algorithm is not within reach. In this work, we demonstrate that the VAT comparison and reordering mechanism can be applied at the level of clusters, instead of its classical application at the level of data items. We exhibit that this innovation results in effective reduction of the computational complexity of the resulting algorithm from $O(N^2)$ to $O(N)$. Moreover, we demonstrate that this technique allows for a generic formulation of the unsupervised clustering problem. This paper includes the mathematical derivation of this idea accompanied by experimental results. We provide some of the deficiencies of the present work at the end of the paper and recommend potential ideas for the continuation of this work.

Chapter 1

Introduction

Clustering algorithms are numerical processes which consume a large number of data items and produce a few cluster representations. The sanity of these processes is commonly controlled by a set of configuration parameters, one of the most important one of which is C, the number of clusters to be found in the data. While many clustering algorithms treat C as an input, sometimes an *unimportant* one, the matter of fact is that properly estimating C is one of the most important tasks that a clustering algorithm must be able to accomplish. In other words, knowing C is an important prerequisite to a usable clustering solution and, except for low data dimensionalities and trivial cluster models, C is not an aspect of the data which can be perceived by *naked eye*. This has been the main driving force behind the development of a vast number of techniques which attempt to perform a robust estimation of C. In the absence of such tools, any clustering algorithm may be prone to generating clustering "solutions" which are seldom usable [1].

Visual Assessment of clustering Tendency (VAT) [2] is, arguably, the rebirth of a group of techniques which have been used in different reincarnations in the past 150 years, this time within the field of data clustering. The core idea behind VAT is that one can estimate the number of clusters present in a set of data items through grouping them based on their mutual distances. In fact, VAT and its many variants have been successfully used in many practical applications within a diverse set of fields of work.

Nevertheless, the model which is at the core of VAT is inherently relational and prototype-based, and is not applicable to many data clustering problems. In fact, as will be discussed later in this paper, the idea that two data items can be compared together, and devoid of their context, is an oversimplification which severely limits the scope of applicability of VAT. Moreover, VAT, in its original form, is not scalable with the size of the input data, and many of its variants inherit this deficiency. Additionally, there are valid arguments in the literature that VAT performs less than ideal as the number and complexity of the clusters increases and for sets of data items which contain outliers and clusters which invade each other's proximity. These conditions are arguably, never an exception, but the norm for data clustering algorithms which are to be used in actual applications which involve real-world data.

Aside from the structural concerns with VAT, this technique is explicitly and inherently based on the subjective opinion of a human user. In other words, VAT is in fact a visualization technique which attempts to, and in a number of cases succeeds to, produce a 2D image which is, supposedly, *easy* to interpret by a human observer. This important concern is in addition to the fact that many VAT-style algorithms require the proper, or sometimes *diligent*, adjustment of one or more configuration parameters.

In this paper, we borrow the core concept of VAT, but instead of applying it on the data items directly, we produce a cluster Dissimilarity Matrix (DM). In fact, we provide theoretical justification to show that the direct application of VAT on the data items is inherently relational, and thus limiting in scope. Then, in order to populate the cluster DM, we utilize an available robustified fuzzy clustering algorithm within a single-cluster framework and feed it with a set of initial cluster representations. Moreover, although we utilize a VAT-sytle reordering process, we discard the Reordered Dissimilarity Matrix (RDM) and use the Minimum Spanning Tree (MST), which is generated during the reordering mechanism. Hence, in direct contrast to the available VAT literature, which uses the RDM and discards the MST, we find the RDM less than useful but find an important application for the MST. There, not only we infer C from the MST, but also we partition the MST into a number of subtrees, each of which yields a cluster representation.

The rest of this paper is organized as follows. First, in Chapter 2, we review the literature of the problem. Then, in Chapter 3, we outline the developed method, and, in Chapter 4, we discuss experimental results. The paper concludes with Chapter 5, which provides the concluding remarks.

Chapter 2

Literature Review

The thinking process behind some of the earliest attempts at the estimation of C, is one of "find many clusters, select only a few" [3]. In this line of thinking, the clustering algorithm is allowed to generate many cluster *candidates* and significant resources have been allocated to the study of cluster validity assessment techniques. The reader is referred to a comprehensive review of this after-the-fact approach in [4] (also see Dunn's index [5], the DB index [6], and the PBM index [7]). Nevertheless, a majority of those approaches make the assumption that the validity of a cluster, or an entire clustering solution, can be measured using a scalar value. However, thorough examination of 23 scalar measures of cluster validity, researchers have shown that *"none* of them are exceptionally reliable across a wide range of datasets" [8] (also see [9] for a Monte Carlo evaluation of 30 different validity indexes). From a theoretical perspective, too, it has been argued that scalar cluster validity indexes aggregate the entire information available in an input set of data items into one or a few metrics and that invaluable information is lost in this process. In the words of the authors of [10], "scalar measures of cluster validity are famously unreliable".

Nevertheless, in comparison between before-the-fact techniques, which estimate the "correct" number of clusters, and after-the-fact approaches, which validate one or a set of clusters, practical implications lean towards the former. In effect, one is inclined, if possible, to execute the clustering process with the proper settings, as opposed to moving ahead with some settings when there is the likelihood that the results are likely to need to be discarded. That situation is most drastic when it is suggested that the clustering algorithm is to be executed for a range of number of clusters in order for the most optimal solution to be picked later.

The findings regarding the inherent deficiencies of scalar clustering validity indexes have encouraged the community to investigate alternative techniques which visualize or assess the inherent structure of a given set of data items from the vantage point of clustering. It is important to emphasize that some of those techniques utilize generic statistical methods and are sometimes expensive to carry out or require marginal probability distributions which are hard or impossible to generate in practical settings.

Visual Assessment of clustering Tendency (VAT) [2] utilizes the pairwise dissimilarity information between data items as a symmetrical matrix with non-negative elements and zero diagonals. We address this matrix as the Dissimilarity Matrix (DM) (this matrix has also been called the Dissimilarity Image (DI)). VAT provides a mechanism for reordering the rows and columns of this matrix in a way that signifies the structure in the data. In short, when VAT is successful, dense areas in the dataset yield dark squares along the diagonal of the Reordered Dissimilarity Matrix (RDM). One can, therefore, at least theoretically, count these squares and generate a reasonable estimate of the number of clusters present in the data. We note that VAT must be reviewed in the context of other visualization techniques such as trees, dendrograms, castles, and icicles [11]. More specifically, VAT belongs to the subset of techniques which utilize image-based visualizations.

An implementation of the core technique of VAT can be found in [12] (for details refer to [13]). The VAT reordering algorithm is based on Prim's algorithm [14] for finding the MST corresponding to a weighted graph (also see [15]). Nevertheless, the intent of VAT is not to generate the tree, but to produce the order in which the vertexes are added as the tree grows. Enhanced VAT (E-VAT) [16] is a variant of VAT which applies a robust loss function on the DM before the reordering process begins in order to limit the impact of the outliers. E-VAT uses Otsu's threshold [17] for the elements of the DM as scale.

The roots of VAT can be traced back to the method known as SHADE [18]. SHADE uses over-striking of printed characters in order to generate a halftone display. Aside from this visualization technique, which is a direct product of the display technologies available at the time of the publication of that paper, SHADE is different from VAT in that it is essentially a cluster visualization method which is utilized after another hierarchical clustering scheme is applied on the data. Additionally, SHADE only generates the lower triangular section of the DM. A variant of SHADE, addressed as the "graphical method of shading" [19], performs quantization of the DM prior to rearrangement. VAT is also related to the Single Linkage (SL) [20] algorithm. SL, in essence, cuts the largest edges in the MST, thus producing subsets of the data, each of which corresponds to an individual cluster. The reader is referred to [21] for a review of the different variations of SL algorithms and to [22] for a review of the "direct relation" [22] between the clusters generated by SL and the reordering prescribed by VAT. The reader is also referred to [23] for a list of practical utilizations of SL as well as a list of clustering algorithms which utilize SL. Nevertheless, it has been argued that SL "famously" [24] fails for some cluster types. Moreover, it has been suggested that the technique of using a shaded matrix for assessing the structure of the data generates a "visual puzzle" which is a "crypto-graphical mystery" [11].

In a broader perspective, the core concept of VAT has been known for many decades and has been utilized in many different contexts. In effect, VAT belongs to the general category known as Cluster Heat Maps [25]. In that paper [25], the authors find traces of the utilization of the display mechanism employed by VAT in late 19th century [26] and a diverse presence in the statistical literature in the 20th century [27]. The reader is referred to [25] for a long list of related references. In fact, Cluster Heat Map techniques are suggested to be "by far the most popular graphical representation" [28]. Moreover, Seriation approaches are among the well-known methods which reorder data items based on some notion of dissimilarity. The reader is referred to [29] for a list of loss functions used in the context of seriation and [30] for a comprehensive review of different Dissimilarity Plots.

While VAT-style algorithms utilize symmetrical square dissimilarity matrices, there are applications in which the rows and the columns of the DM refer to different mathematical objects. An extension of VAT, nicknamed coVAT [31], has been developed for this category of problems. coVAT essentially treats the rectangular DM as the known section of a larger square DM and then *imputes* the missing values using a transitive logic. That technique is extended in further works, including Scalable co-VAT (scoVAT) [32], coVAT2 [33] and Improved co-VAT (co-iVAT) [34].

Review of the artificial samples utilized in a number of works which involve VAT (see for example [2, 10, 24, 35]) indicates that there are no outliers in these sets of input data items. In fact, VAT is fundamentally based on the assumption that the input set of data items is composed of a number of compact well-separated isotropic clusters. As stated by the authors of [36], VAT is "usually only effective at highlighting cluster tendency in data sets that contain compact well-separated clusters". However, as also highlighted by the authors of the same paper, "[m]any practical applications involve data sets with highly irregular structure, which invalidate this assumption" [36]. We argue that the need for working with what is sometimes called "irregular geometries" [36] or "composite shape(s)" [37] is an important practical requirement. In fact, we suggest that what is often addressed as "tough cases" [34] is in fact *the* situation that needs to be addressed by a clustering algorithm. These are cases, for which, in words of the authors of [34], "VAT fails to accurately show the cluster tendency".

Another one of the many successors of VAT is the Visual Cluster Validity (VCV) approach outlined in [10]. In fact, VCV inherits from and build upon both SHADE and VAT and contributes to them the utilization of a generic cluster model. This is as opposed to the assumption by VAT that a cluster can be represented by a typical data item. Nevertheless, VCV, like SHADE, is a tool which displays the clusters that have been produced by another "outsourced" clustering algorithm. In other words, VCV transforms clusters, that are potentially defined in a hyper-dimensional geometry, into representations which can be visualized using a 2D matrix. Nevertheless, the way VCV approaches this capability is through cluster-to-cluster comparison, which is carried out based on the Euclidean distance between the parameters which define the cluster model. Hence, for example, in the context of an ellipsoidal geometry, VCV will combine the components of the mean vector with the elements of the covariance matrix. We argue that this approach mixes into a pot elements which not only accept values at very different scales, but also, and more importantly, are different from a theoretical perspective. Visual Cluster Validity (VCV2) [38] is a similar approach which uses image matching in order to compare cluster representations with the RDM.

Nevertheless, VCV utilizes the context of the data items when it estimates the dissimilarity between a pair of data items. In order to do so, VCV models the dissimilarity between two data items as the minimum of their common distance to a set of clusters. This process provides an important contribution over the simplified models which utilize explicit data item-to-data item distances as their level of dissimilarity, as, for example, is carried out by VAT and a majority of its variants. However, assessment of the performance of VCV for Euclidean and linear clusters in $\mathbb{R}^2$ in [10] shows that the results corresponding to linear clusters "are not nearly as clean as" [10] those for Euclidean clusters. Additionally, VCV employs an overestimated number of clusters in order to recognize the structure of the data and "some deterioration" [10] is observed when the clustering algorithms is executed using cluster counts which are "large" [10]. We note that the problem classes utilized in [10] are the equivalents of the 2de and 2dl models employed in this paper. We also emphasize that the sets of data items which are utilized in [10] appear to not contain outlier data items.

Revised VAT (reVAT) [39] replaces the RDM with a set of profile graphs. That technique allows for removing data items which appear to be highly likely to belong to the same cluster from the pool of data and has been called "quasi-ordering" as well. That naming convention is in contrast with the full reordering carried out in VAT. Nevertheless, the output of reVAT is of a different form, compared to VAT, and, therefore, as stated by the developers of reVAT, "interpretation of the set of reVAT profile graphs becomes very difficult when the number of clusters is large, or there is significant overlap between groups of objects in the data" [24]. In fact, it has been argued [24] that reVAT is applicable when $C \leq 5$. This deficiency is alleviated in bigVAT [24], in which the visual form of the VAT output is re-instituted, but the DM is confined to samples which are proportionally selected from the dominant reVAT profiles.

Another sequel to VAT is Scalable VAT (sVAT) [35]. The development of sVAT is the result of the realization that although bigVAT produces a 2D visualization, but that "this image may not be as descriptive as a VAT-ordered image" [35]. Nevertheless, sVAT, too, performs sampling on the data in order to achieve a lower computational complexity. In a similar fashion to bigVAT, sVAT also utilizes an estimate of the clusters present in the input set of data items in order to produce a representative subset. However, in addition to the requirement that the full DM is provided by the user, sVAT relies on the user for providing the "desired (approximate) sample size" and an "overestimate of the true number of clusters". An extension of sVAT, nicknamed sVAT-SL, in which SL stands for Single Linkage, is suggested in [40]. sVAT-SL attempts to produce both the cluster representations as well as C. Nevertheless, sVAT-SL is in fact interrupted at the middle of the process, because "the user must choose the number of clusters ... to seek" [40]. In other words, the user of sVAT-SL is requested to observe the RDM and to decide what C ought to be. Nevertheless, while sVAT-SL is advertised as "an approximation to single-linkage clustering for big data" [40], the same paper also asserts that the clustering results generated by sVAT-SL "can be ruined by outliers" [40].

It has been argued that "a major limitation" [41] of VAT and its variants is their "inability to highlight cluster structure ... when ... [the data] contains clusters with highly complex structure" [41]. Spectral VAT (SpecVAT) [41] attempts to increase the legibility of the RDM generated by VAT algorithms through spectral decomposition of the DM prior to the reordering. However, the performance of SpecVAT depends on the proper selection of the parameter k, i.e. the number of eigenvectors used during decomposition. In fact, as demonstrated in [36], to properly select k, one ought to have a proper estimate for C, the number of clusters present in the data. The critical necessity for the existence of this *a priori* piece of information violates the premise behind VAT and SpecVAT. The latter work utilizes image processing techniques on the RDM and employs a sampling framework in order to reduce the computational complexity of the resulting algorithm. It also attempts several ideas for estimating the parameters which govern the process in order to resolve some of the concerns with the original SpecVAT.

In spite of their differences, VAT, reVAT, bigVAT, sVAT, and many other algorithms in their class, are in fact *visual assessment* methods. In other words, these techniques are inherently reliant on the *subjective* [24] understanding of the user. This issue becomes more disconcerting when it is argued that in some practical settings, an "experienced user" [24] may have to be employed in order to perform the assessment (the reader is referred to [37] for related remarks). In fact, it has been argued that non-Euclidean geometries or overlap between clusters can give rise to VAT images for which "different viewers may deduce different numbers of clusters ..., or worse, not be able to estimate c at all" [36]. To make matters more complicated, the output of many of the algorithms in this class is a two dimensional image, which requires to be transferred and displayed *diligently* and it is argued that compression, down-scaling, and interleaving of this image "may obscure important information about potential clusters in the data" [24].

Another important detail about VAT and a majority of its affiliates is their treatment of the MST. In fact, in many of the algorithms in this category, the MST is in essence considered as a by-product which is discarded as soon as it is created. We argue, however, that this representation carries very important information about the structure of the data and the relationship between the clusters present in it. As will be shown later in this paper, we argue that the RDM is an alternative, and less useful, representation than the MST. Nevertheless, significant effort has been spent in the literature on the interpretation of the structure of the data based on the appearance of the RDM. Cluster Count Extraction (CCE) [42] is one such approach, which assesses the histogram of the DM using image processing operators such as Otsu's algorithm [17] and Fast Fourier Transform (FFT). CCE requires the proper adjustment of multiple threshold and filter radius parameters. Dark Block Extraction (DBE) [43] is another approach which counts the number of dark diagonal blocks in the RDM using thresholding and morphological operations. DBE also requires the diligent adjustment of the threshold value used for determining the peaks of a histogram. aVAT [36] is another approach which utilizes image processing techniques in order to interpret the RDM. That paper applies a function, which can be considered a robust loss function, on the RDM in order to increase its contrast and to facilitate the recognition process. The reader is also referred to [44] in which Dunn's cluster validity index is used in order to perform thresholding on the RDM. Other validity indexes, such as the PBM index [7], have been used for interpreting the RDM as well. Clustering in Ordered Dissimilarity Data (CLODD) [45] attempts to automatically produce the clusters from the RDM as well. While it is advertised that "when [the input data] has "good" clusters, CLODD will find them", the performance of CLODD depends on the values of the two "influence constant"s α and γ, the values of which are set using trial and error [45].

It is important to emphasize that, independent of the actual efficacy of the available approaches which utilize image processing tools in order to interpret the RDM, those techniques are in essence based on the assumption that the RDM is *the* representation to be processed and that the MST is *inferior* to it. This is evident in the concluding statement in [36], in which the authors advocate for the use of more efficient thresholding techniques in order to enhance the interpretation of the RDM. We argue, however, that a more important consideration in this discussion is the choice of representations, and that the MST is a *more* informative representation than the RDM.

For an input set of data items $\mathbf{X}$, which contains N data items, the computational complexity of VAT is of $O(N^2)$. This is due to the fact that VAT processes the $N \times N$ matrix of dissimilarities between the data items. In addition to the fact that this model is inappropriate in the context of a generic notion of homogeneity, the quadratic complexity of VAT is prohibitive for Big Data problem instances. In fact, it has been argued that VAT "works well for relatively small data sets ($n \leq 500$)" [24]. Hence, variations of VAT have been proposed which address this issue. reVAT, bigVAT, and sVAT are three approaches which aim at lowering the computational complexity to $O(CN)$. Here, C is the inherent, and *unknown*, number of clusters which are present in $\mathbf{X}$. As commonly $C \ll N$, this transformation is greatly beneficial. Nevertheless, it has been argued that "sVAT does not asymptotically scale linearly with [the number of data items]" [40].

The $O(N^2)$ computational coomplexity of VAT contains the superpositions of the costs associated with two processes, both of which require $O(N^2)$ operations. In fact, not only the DM needs to be reordered at the cost of $O(N^2)$ operations, but also this matrix needs to be calculated in the first place, and the computational complexity of that process is of $O(N^2)$ as well. In this context, the sequels of VAT, i.e. reVAT, bigVAT, and sVAT, drop the complexity of the reordering mechanism to $O(CN)$, but, nevertheless, they still require the calculation of the $N \times N$ DM. Hence, technically, the computational complexities of these algorithms are still $O(N^2)$, unless the DM is provided *a priori*. As will be discussed next, that circumstance is only applicable to a small subset of clustering problems, in which case, too, the DM must be calculated in many practical settings anyways. Hence, neither VAT nor the aforementioned variants of it are *genuinely* scalable. Here, we rely on the notion of scalability which mandates that the computational complexity of the algorithm must grow linearly as the number of input data items increases [46].

Nevertheless, a critical concern with VAT, and other models in its category, is in their core assumption about the nature of the inter-relationship between the data items. In fact, VAT is confined to situations in which it is meaningful to discuss the similarity/dissimilarity of two data items in vacuum. In this context, VAT makes the extremely limiting assumption that the data items are either given as object vectors or by numerical pairwise dissimilarity values [2]. Hence, VAT assumes that given the two data items x_1 and x_2, one can meaningfully point out the extent to which the two are likely to belong to the same cluster. This assumption is valid within the context of the Euclidean distance function as well as other ℓ_k norms in which case the clusters are in essence prototypical data items.

In fact, in the literature, the input set of data items has been, *incorrectly* in our opinion, reduced to and assumed to be equivalent to its relational representation (for recent examples refer to [22, 36]). This limited scope is more obvious in the case of algorithms such as Relational Visual Cluster Validity (RVCV) [47] and Correlation Cluster Validity (CCV) [48], which explicitly identify their narrow scope. While that situation is applicable to relational and prototype-based clustering problem classes, it is not necessarily an inclusive framework. We argue that the notion of "dissimilarity between two data items" may not be meaningful for many problem classes which are highly relevant in practice, unless a context is defined. In other words, the question is *not* how dissimilar the two data items x_1 and x_2 are. One example to illuminate this point is clustering members of a given $\mathbb{R}^k$ into lower dimensional spaces. Under this regime, if data items are to be clustered into lines, planes, and hyperplanes, for example, the question of dissimilarity between two data items is *meaningless* unless one also provides a cluster representation. In other words, any pair of data items x_1 and x_2 may be extremely similar or inherently dissimilar given their mutual relationship to the cluster which provides the context. Hence, we argue, the square DM representation is only applicable to a small subset of possible problem classes. The reader is referred to SpecVAT [41], as only one example of, the "natural" and implicit reduction of a set of data items to its relational representation.

VAT approaches have been augmented with path-based distance models in order to alleviate the limiting scope mandated by the assumption of the relational model. For example, Improved VAT (iVAT) [36] prescribes that it is not the direct data item-to-data item distance which must be stated in the DM but that x_1 and x_2 are "similar" if there is a sequence of data items, with x_1 and x_2 at the two ends of the sequence, each of which are at close distances to each other.

Sample experimental results provided in [36] suggests that iVAT is capable of recognizing arbitrary sequences of data items in $\mathbb{R}^2$. The authors of that paper also propose the Automatic VAT (aVAT) technique which applies a function, that can be considered a robust loss function, on the RDM in order to facilitate the semi-unsupervised extraction of the diagonal blocks from it. The reader is also referred to a variant of the iVAT algorithm, nicknamed Efficient iVAT (efiVAT), which utilizes dynamic programming [34]. A related algorithm, named clusiVAT [23], samples the input set of data items in order to generate the cluster representations using SL. clusiVAT then extends the classification results to the entire set of data items. The reader is referred to [49] for variants of clusiVAT and iVAT, nicknamed clusiVAT+ and iVAT+, which utilize additional "efficient thresholding schemes". While these contributions provide some relief, especially when the clusters are isolated and separated, the core problem is still far from being addressed by these techniques. One scenario in which the path-based model falls short of resolving the challenge is the clustering of members of $\mathbb{R}^k$ into lower-dimensional spaces. In this scenario, any intersection between a pair of clusters is a real threat that may cause the two clusters to "leak" into each other. This case is closely related to the "zigzagging" phenomenon which the original implementation of VAT was prone to and was alleviated using a clever initialization procedure for the reordering mechanism [2].

Moreover, even the mere assumption that the data items can necessarily be reduced to members of $\mathbb{R}^k$ is in essence a dangerous reduction. As will be shown later in this paper, the data items merely need to be *mathematical objects* for which *distance to a given cluster* can be defined. This is a major step forward compared to the majority of the works in the VAT literature, where the assumption that $x_n \in \mathbb{R}^k$ is made out of convenience, without any reference to the implications of that reduction.

There is a significant overlap between the group of individuals who worked on VAT and Fuzzy C-Means (FCM) [50]. While, historically, FCM precedes VAT, VAT is in fact an attempt to set one of the key inputs required by FCM, i.e. the value of C. Additionally, as stated above, there have been multiple attempt to forge VAT into an alternative to FCM, i.e. to generate the cluster representations, and not just their count, by VAT. This convoluted relationship between VAT and FCM, however, to the best of our knowledge, has never led to a marriage of the two ideas. In other words, VAT either feeds into FCM or replaces it, but never intertwines with it. In this paper, we take up this task and demonstrate that a robustified variant of FCM, nicknamed *Connie* [51], can be used as a single-cluster clustering mechanism, i.e. by assuming $C = 1$, in order to scan the cluster space and to produce cluster representations which are aggregated using a VAT reordering process in order to produce a MST, the sub-trees of which yield the sought for clusters in the input set of data items.

Chapter 3

Method

3.1 Modeling Framework

Any unsupervised data clustering problem is based on a *notion of homogeneity* which is rooted in the physical properties of the data items and the clusters. Therefore, there is an important practical incentive for discussing unsupervised data clustering in generic terms and independent of the framework of any particular *problem class*. Here, a problem class is the mathematical formalization of the data clustering problem in the context of a particular model for the data items and a particular notion of homogeneity. This notion of homogeneity contains a cluster model. For example, one may refer to the problem class which relates to "Euclidean clustering of points which belong to $\mathbb{R}^2$". In this statement, "Euclidean clustering" defines the notion of homogeneity, which then mandates that each cluster is represented as a point in $\mathbb{R}^2$, thus defining the cluster model relevant to the problem class in hand. Additionally, the clause "points which belong to $\mathbb{R}^2$" provides the model for the data items. Once a problem class has been defined, one can discuss particular *problem instance*s. Here, a problem instance is one realization of a particular problem class. In other words, a problem instance provides a set of data items, as prescribed by a problem class.

Under such circumstances, the vision of this work is to develop a generic fuzzy clustering algorithm which accepts data item and cluster models as plug-ins and operates using a data item-to-cluster distance function which is provided as a black-box. In other words, we propose a generic unsupervised fuzzy clustering algorithm which can be adopted to any problem class. Once the adoption is carried out, this particular incarnation of the proposed method will operate on any instance of the aforementioned problem class without any need for user supervision or subjective intervention.

3.2 Model Preliminaries

As discussed in Section 3.1, any problem class provides a mathematical model for the data items. Additionally, any problem class defines a cluster model which complies with the notion of homogeneity relevant to the problem class at hand. We denote a data item as x and a cluster as ψ.

In this work, we utilize a weighted set of data items, defined as,

$$\mathbf{X} = \left\{ (\omega_n; x_n) \right\}, n = 1, \cdots, N, \omega_n > 0, \tag{3.1}$$

and we define the *weight* of $\mathbf{X}$ as,

$$\Omega(\mathbf{X}) = \sum_{n=1}^{N} \omega_n. \tag{3.2}$$

The notion of associating positive weights to the data items can be considered as a marginal case of clustering fuzzy data [52]. Examples for this setting include clustering of a set which is inherently weighted, clustering of sampled data, clustering in the presence of multiple classes of data items with different priorities [53], and a measure used in order to speed up the execution through data reduction [54]. When known in the context, we abbreviate $\Omega(\mathbf{X})$ as Ω. Thus, when estimating expected values, we treat $\mathbf{X}$ as a set of realizations of the random variable x and write,

$$p\{x_n\} = \frac{\omega_n}{\Omega}. \tag{3.3}$$

We model the relationship between a data item and a cluster as the real-valued positive *distance function* $\phi(x, \psi)$. Through this abstraction, we decidedly avoid the dependence of the underlying algorithm on Euclidean or any other particular notations of distance. We also assume that the distance function is unbounded, i.e. that for any cluster representation ψ and any positive value L, there exist infinite number of data items x for which $\phi(x, \psi) > L$.

We assume that the robust loss function, $u(\cdot) : [0, \infty] \to [0, 1]$, is given which satisfies $\lim_{\tau \to \infty} u(\tau) = 1$. Additionally, we assume that $u(\cdot)$ is an increasing differentiable function which satisfies $u(0) = 0$ and $u(1) = \frac{1}{2}$. Hence, in this work, we utilize the rational robust loss function,

$$u(x) = \frac{x}{1+x}, \tag{3.4}$$

and we model the loss of x_n when it belongs to ψ_c as,

$$u_{nc} = u\left(\frac{1}{\lambda}\phi_{nc}\right), \phi_{nc} = \phi(x_n, \psi_c). \tag{3.5}$$

We model the loss of a data item which is considered to be an outlier as the positive constant U. In (3.5), we address λ as the *scale* parameter (note the similarity with the cluster-specific weights in PCM [55]). In fact, λ has a similar role to that of scale in robust statistics (also called the *resolution parameter* [56]) and the idea of distance to noise prototype in the NC algorithm [57, 58]. Scale can also be considered as the controller of the boundary between inliers and outliers [59]. From a geometrical perspective, λ controls the radius of spherical clusters and the thickness of planar and shell clusters [60].

We assume that $\phi(x,\psi)$ is differentiable in terms of ψ and that for any non-empty weighted set **X**, the following function of ψ,

$$\Delta_{\mathbf{X}}(\psi) = E\{\phi(x,\psi)\} = \frac{1}{\Omega}\sum_{n=1}^{N}\omega_n\phi(x_n,\psi), \qquad (3.6)$$

has one and only one minimizer which is also the only solution to the following equation,

$$\sum_{n=1}^{N}\omega_n\frac{\partial}{\partial\psi}\phi(x_n,\psi) = 0. \qquad (3.7)$$

In this paper, we assume that a function $\Psi(\cdot)$ is given, which, for the input weighted set **X**, produces the optimal ψ which minimizes (3.6) and is the solution to (3.7). We address $\Psi(\cdot)$ as the *cluster fitting function*.

Note that $\Psi(\cdot)$ is the solution to the M-estimator given in (3.6). We emphasize that when a closed-form representation for $\Psi(\cdot)$ is not available, conversion to a W-estimator can produce a procedural solution to (3.7) [61]. Additionally, many of the techniques developed in the context of Weber Problems [62] may be applicable to finding a procedural solution to $\Psi(\cdot)$.

A version of this modeling framework has been called a "prototype generator" [10] in that a cluster is modeled using a finite set of mathematical entities which *generate* the set of data items that belong to the cluster. The entirety of this model has precedence in the literature and has been used in parallel [51] as well as sequential settings [63].

3.3 Assessment of Loss

In this section, we carry the loss modeling framework developed in [51] in order to render a self-containing paper. This framework derives a loss model for the set **X** which is *known* to contain C clusters. We then derive the single-cluster version of this loss model.

We assume that, at some arbitrary point during the procedure, a clustering algorithm has discovered the C clusters $\psi_1,\cdots,\psi_C$ in **X**. We also assume that a Maximum Likelihood procedure has been applied on $\mathbf{X}$ and denote the set of data items which are assigned to ψ_c as $\tilde{\mathbf{X}}_c$. We address the union of all $\tilde{\mathbf{X}}_c$ for $c = 1, \cdots, C$ as $\tilde{\mathbf{X}}$. In this context, the set $\tilde{\mathbf{X}}_0 = \mathbf{X} - \tilde{\mathbf{X}}$ contains the data items which are considered to be outliers.

Now, we consider an arbitrary data item x_n. This data item may be an outlier or it may belong to one of the C clusters. Hence, we model the loss associated with x_n as follow.

$$E\{Loss|x_n\} = p\{x_n \in \tilde{\mathbf{X}}_0\} E\{Loss|x_n \in \tilde{\mathbf{X}}_0\} + \qquad (3.8)$$
$$p\{x_n \in \tilde{\mathbf{X}}\} \sum_{c=1}^{C} p\{x_n \in \tilde{\mathbf{X}}_c | x_n \in \tilde{\mathbf{X}}\}$$
$$E\{Loss|x_n \in \tilde{\mathbf{X}}_c\}.$$

We now denote the probability that x_n is an inlier as p_n and the probability that x_n belongs to $\tilde{\mathbf{X}}_c$, given that it is an inlier, as f_{nc} and rewrite (3.8) as follows.

$$E\{Loss|x_n\} = (1 - p_n)U + p_n \sum_{c=1}^{C} f_{nc} u_{nc}, \qquad (3.9)$$

where (3.5) is used. Note that the definition of f_{nc} mandates that,

$$\sum_{c=1}^{C} f_{nc} = 1, \forall n. \qquad (3.10)$$

Then, we aggregate (3.8) for all x_n and use (3.3) and write,

$$E\{Loss|\mathbf{X}\} = \sum_{n=1}^{N} p\{x_n\} E\{Loss|x_n\} = \qquad (3.11)$$
$$\frac{1}{\Omega} \sum_{n=1}^{N} \omega_n \left[p_n \sum_{c=1}^{C} f_{nc} u_{nc} + UC \frac{1}{C}(1 - p_n) \right].$$

Close assessment of (3.11) shows that this cost function complies with an HCM-style hard template. It is known, however, that the utilization of the concept of the *fuzzifier* has important benefits (we will get back to this concept later). Hence, we rewrite (3.11) and derive the following cost function.

$$\Delta = \sum_{n=1}^{N} \omega_n \left[p_n^m \sum_{c=1}^{C} f_{nc}^m u_{nc} + UC^{1-m}(1 - p_n)^m \right]. \qquad (3.12)$$

This objective function is to be minimized subject to (3.10).

In (3.12), $m > 1$ is the *fuzzifier* (also called *weighing exponent* and *fuzziness*). The optimal choice for the value of the fuzzifier is a debated matter [64] and is suggested to be "an open question" [65]. It has been suggested that $1 < m < 5$ [66] and $1.5 < m < 2.5$ [67] are proper ranges and that $m = 2$ is an appropriate value [66]. The use of $m = 2$ is suggested in early work on the topic [5, 68, 69] and there is physical evidence for this choice as well [70]. Nevertheless, other researchers [65] have argued that the choices for the value of m are mainly empirical and lack a theoretical basis. The reader is referred to [71] for a review of the concept of fuzzifier and the alternatives for it. In this work, we use $m = 2$ in order to comply with the history of the problem.

In (3.11) we have chosen to write U as $UC\frac{1}{C}$ in order to compensate for the impact of the fuzzifier. In fact, two types of terms exist in (3.11), i.e. $p_n f_{nc}$ and $(1 - p_n)$. These terms are each products of membership identifiers. However, while the first term contains two elements, the second one only contains the single element $1 - p_n$. We argue that this is because this term in fact contains implicit components of the type "*if either P or not P*" hidden in it. In other words, the cost component $U(1-p_n)$ in fact models the situation in which x_n is an outlier, in which case it is irrelevant whether or not x_n belongs to any of the clusters. In other words, the term $(1 - p_n)$ is in fact the simplified version of the following term,

$$(1-p_n)\sum_{c=1}^{C} f_{nc} = 1 - p_n. \qquad (3.13)$$

While this alternative form is in effect identical to $1 - p_n$, the difference becomes significant when the fuzzifier is integrated into the objective function. In fact, with the addition of the fuzzifier, the term given in (3.13) ought to be modified to,

$$(1-p_n)^m \sum_{c=1}^{C} f_{nc}^m \leq (1-p_n)^m. \qquad (3.14)$$

In other words, if no other measure is taken, the incorporation of the fuzzifier effectively reduces the cost of being an outlier, as explained below.

We note that for any set of C non-negative variables ζ_c which satisfy $\sum_{c=1}^{C} \zeta_c = 1$, we have $\sum_{c=1}^{C} \zeta_c^m \leq 1$, when $m > 1$. The equality in this relationship, i.e. the upper bound, occurs when all of the ζ_c are zero except for one which is unity. The lower bound on $\sum_{c=1}^{C} \zeta_c^m$, however, occurs when the ζ_c are identical. Therefore, we replace (3.13) with the case in which all the f_{nc} are equal. This process guarantees that when the fuzzifier is incorporated into the cost function, the corresponding term is always greater than or equal to the pre-fuzzifier term. In other words, we replace $(1-p_n)$ with $(1-p_n)C\frac{1}{C}$ and, therefore, after the incorporation of the fuzzifier, yield $(1-p_n)^m C\frac{1}{C^m} = (1-p_n)^m C^{1-m}$. In the above, this transformation was rephrased, *imprecisely*, as substituting U with $UC\frac{1}{C}$ in (3.11).

3.4 Single-Cluster Clustering

In this section, we set $C = 1$ and derive the single-cluster version of the loss model which was developed in Section 3.3. We then provide a solution strategy for it.

When $C = 1$, the constraint (3.10) reduces the C-element set $\{f_{n1}, \cdots, f_{nC}\}$ to the set $\{1\}$. We also conveniently drop the subscript c from u_{nc} and, therefore, rewrite (3.12) as,

$$\Delta = \sum_{n=1}^{N} \omega_n \left[p_n^2 u_n + (1 - p_n)^2 U \right]. \tag{3.15}$$

Note that Δ is to be minimized void of any constraint.

Calculating $\frac{\partial \Delta}{\partial p_n}$ and equating it to zero, we derive the optimal p_n as,

$$p_n = \frac{1}{1 + \frac{1}{U} u_n}. \tag{3.16}$$

Now, we plug (3.16) in (3.15) and write,

$$\Delta = \sum_{n=1}^{N} \omega_n \frac{U u_n}{U + u_n}. \tag{3.17}$$

Derivation shows that (3.17) can be rewritten as follows, while it is important to emphasize that the right side of (3.18) is to be *maximized*.

$$\Delta \equiv \sum_{n=1}^{N} \omega_n \frac{1}{1 + \frac{1}{U} u_n}. \tag{3.18}$$

Comparison of (3.18) and (3.16) reveals that,

$$\Delta \equiv \sum_{n=1}^{N} \omega_n p_n. \tag{3.19}$$

We then derive,

$$\frac{\partial \Delta}{\partial \psi} = \sum_{n=1}^{N} \omega_n p_n^2 \frac{1}{\lambda} u' \left(\frac{1}{\lambda} \phi_n \right) \frac{\partial}{\partial \psi} \phi(x_n, \psi). \tag{3.20}$$

Using (3.7) we know that the solution to (3.20) is given as,

$$\psi = \Psi \left(\left\{ (\tilde{\omega}_n; x_n) \right\} \right). \tag{3.21}$$

Here,

$$\tilde{\omega}_n = \frac{\omega_n p_n^2}{\left(1 + \frac{1}{\lambda} \phi_n \right)^2}, \tag{3.22}$$

which can be shown to be equal to,

$$\tilde{\omega}_n = \omega_n \left(1 - u_n\right)^2 p_n^2. \tag{3.23}$$

Here, we carry a brief overview of the single-cluster clustering procedure developed in this paper. We note that this process is in fact a Picard iteration.

- **Inputs**

    (a) Cluster representation ψ.

    (b) Input set of data items **X**.

- **Outputs**

    (a) Updated cluster representation ψ.

    (b) Membership identifiers $p_n, n = 1, \cdots, N$.

- **Procedure**

    (a) Set $p_n = 1, n = 1, \cdots, N$.

    (b) Loop

        (a) Calculate ϕ_n and u_n for $n = 1, \cdots, N$, using (3.5).

        (b) Calculate $\tilde{\omega}_n$ for $n = 1, \cdots, N$, using (3.23).

        (c) Calculate ψ using (3.21).

        (d) Calculate Δ using (3.19).

        (e) If change in Δ is negligible, break the loop.

    (c) Calculate p_n for $n = 1, \cdots, N$, using (3.16).

3.5 Cluster Space Sampling

The process carried at the end of Section 3.4 inputs a cluster representation and modifies it into a *more optimal* representation. In effect, this process accepts a point in the cluster space and updates it through a local search mechanism. The fact that that process functions locally, is an important asset, because it can ignore the rest of the data items and only "focus" on a locality of **X**. Nevertheless, this same phenomenon means that it is probable that ψ would in fact correspond to a local minimum of the search space and that there is no guarantee that in every execution of that process, using any input ψ, the generated cluster in fact describes one of the major homogenous sets present in the input set of data items.

Additionally, it is important to discuss the mechanism through which ψ is generated. Here, we denote three potential sources for ψ, i.e. cluster space sweeps, *a priori* pieces of information, and random cluster representations. We describe these three mechanisms in detail in the next paragraphs. Each one of these processes generates a cluster population, which we denote as $\boldsymbol{\Psi}$. Note that, the number of elements in this set is independent of C and is determined based on the available budget for processing power and also model complexity.

First, $\boldsymbol{\Psi}$ may be generated by a process which "sweeps" the cluster space, given that the cluster model allows for such an action. For example, in one imaginary situation, the cluster representation may contain a point in a rectangular subset, for example $[-L, L] \times [-L, L]$, and an angle. In such circumstances, having made a decision about the size of $\boldsymbol{\Psi}$, one can devise step sizes along the three dimensions of the cluster space and then to uniformly sample this space. The sweeping mechanism provides assurance that the entire cluster space is examined. Nevertheless, designing a sweeping mechanism for an arbitrary cluster representation may be non-trivial and cumbersome.

A second approach to producing $\boldsymbol{\Psi}$ is to utilize *a priori* pieces of information. For example, due to the inherent properties of the problem class, it may be known that clusters tend to agglomerate in certain "areas" in the cluster space. If this condition is applicable to a problem class, then one may capitalize on the opportunity and generate $\boldsymbol{\Psi}$ as the concatenation of cluster representations located in these "hot spots".

The third approach, which is also the one adopted in this paper, is to populate $\boldsymbol{\Psi}$ with random cluster representations which are generated using i.i.d. processes. This approach is easy to implement, especially when ψ is in fact a vector in a hyper-rectangular subspace of some $\mathbb{R}^k$.

We assume that one of the above mechanisms, or any other applicable mechanism, has been employed and that the set $\boldsymbol{\Psi}$, which contains $\hat{C}$ cluster representations is generated. Here, $\hat{C}$ is significantly larger than any estimate for C. We feed these cluster representations to the algorithm described in Section 3.4 and generate the updated set $\hat{\boldsymbol{\Psi}}$ as well as $\hat{C}$ instance of $\{p_1, \cdots, p_N\}$ identifiers. Note that the computational complexity of this process is of $O(N\hat{C})$, wherein $\hat{C}$ is independent of both C and N. In other words, the computational complexity of the method developed in this paper scales linearly with the size of the input set of data items and, hence, this method is *scalable* [46].

3.6 Cluster Aggregation

We utilize VAT in order to extract homogenous sets from $\hat{\Psi}$. Each one of these sets contains one group of clusters which are *similar*. In order to be able to compare clusters, we utilize a Bayesian risk-based comparison framework. In other words, the two cluster representations ψ_1 and ψ_2, are *dissimilar* to the extent that they produce contradicting classification results. In other words,

$$\delta_{\psi_{c_1},\psi_{c_2}} = E\left\{p\{misclassification\}\right\} = \tag{3.24}$$

$$\max\left[E\left\{p\{x_n \notin \mathbf{X}(\psi_{c_2})|x_n \in \mathbf{X}(\psi_{c_1})\}\right\},\right.$$

$$\left. E\left\{p\{x_n \notin \mathbf{X}(\psi_{c_1})|x_n \in \mathbf{X}(\psi_{c_2})\}\right\}\right] =$$

$$\max\left[\frac{\sum_{n=1}^{N}\left[p_{nc_1} > \frac{1}{2}\right]\left[p_{nc_2} < \frac{1}{2}\right]\omega_n}{\sum_{n=1}^{N}\left[p_{nc_1} > \frac{1}{2}\right]\omega_n},\right.$$

$$\left.\frac{\sum_{n=1}^{N}\left[p_{nc_1} < \frac{1}{2}\right]\left[p_{nc_2} > \frac{1}{2}\right]\omega_n}{\sum_{n=1}^{N}\left[p_{nc_2} > \frac{1}{2}\right]\omega_n}\right].$$

Here, $\mathbf{X}(\psi)$ denotes the set of data items which are classified to the cluster represented by ψ and $[p]$ is the *Iverson bracket*, wherein $[p]$ is one(zero) if the Boolean variable p is true(false). For simplicity, we denote $\delta_{\psi_{c_1},\psi_{c_2}}$ as δ_{c_1,c_2}. We envision that as p_{n_1} and p_{n_2} denote probabilities, one may be able to utilize other probabilistic distance metrics [72].

Having produced δ_{c_1,c_2} for all pairs of $1 \leq c_1 < c_2 \leq \hat{C}$, we arrive at a cluster DM which we process using the classical VAT reordering algorithm. Note that the computational complexity of generating the DM in the algorithm developed in this paper is of $O(\hat{C}^2)$. This number, too, is independent of N. Moreover, as stated before as well, we do not have any particular use for the RDM which is generated through applying VAT on the DM. In fact, we perform VAT on cluster DM for the sole purpose of generating the MST. We represent this graph as the sequence of element-to-element dissimilarities as the reordering mechanism proceeds. Hence, one may in fact rewrite the VAT algorithm so that the RDM is not generated at all.

We cut the longest links in the MST using the preset threshold of 5%. In other words, we consider two clusters to be distinct when the probability of nonconforming classification between them is more than 0.05. We also mandate that every cluster group must at least contain 5% of $\hat{\Psi}$ and denote this number as the "prominence" of the corresponding cluster. We note that this metric is distantly related to the size of the dark blocks in the RDM in classical VAT and many of its counterparts. Nevertheless, and this is very important to emphasize, the RDM blocks in the present work, although the RDM is never generated, represent estimates for the probability that a cluster may converge to any of the sub-trees. The RDM blocks in VAT, bigVAT, and other algorithms in that class, however, denote an estimate of the number of data items which are "similar" to each other.

The division of the MST into multiple sub-trees has precedence in the literature. In [40], the authors first estimate C and then cut the $C-1$ longest links in the MST. On the contrary, in this work, we examine long links in the MST in order to both, estimate C, as well as to generate the sets of similar clusters. Then, we utilize the first cluster representation in each sub-tree as the representative of that sub-tree. The reader is encouraged to utilize more optimal, and therefore more costly, approaches. Nevertheless, we also calculate the "weight" of each clusters. Here, weight of a cluster is the relative number of data items which are classified into that particular cluster. We do not utilize weight in order to validate a cluster.

3.7 Determination of U and λ

The notion that a fuzzy clustering algorithm functions within an *unsupervised* framework, is often misunderstood and inaccurately stated. In effect, unsupervised clustering is commonly defined in contrast with *supervised* clustering, wherein a *supervisor* provides training data and possibly direction and guidance in order for the algorithm to function desirably. Unsupervised clustering, on the other hand, is the term used for the broad category of algorithms which *theoretically* do not require user supervision. Nevertheless, as pointed in reference to numerous pieces of work in the literature in Chapter 2, many available "unsupervised" clustering algorithm depends on the diligent adjustment of one or several configuration parameters. In many cases, these parameters are to be adjusted using trial-and-error or "user knowledge". We emphasize, however, that the only type of configuration variables acceptable to be employed by a *truly* unsupervised algorithm are those for which a deterministic and objective selection procedure is outlined which allows the user to be agnostic of the internal mechanics of the algorithm under consideration.

The execution of the algorithm developed in this paper is governed through the two configuration

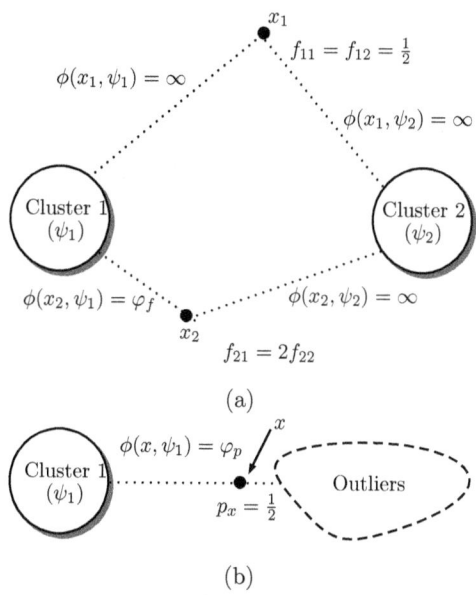

Figure 3.1: The process of determining λ and U. (a) Selection of φ_f. (b) Selection of φ_p.

parameters λ and U. In this section, we provide objective procedures for determining the values of these two configuration variables for any arbitrary problem class. It is worth emphasizing that λ and U are to be set at the level of problem classes and that they are not affected by the particular set of input data items associated with any problem instance or the number of clusters present in it.

It is evident that λ defines the scale for ϕ_{nc}. This is exemplified in (3.5) and also everywhere else in this paper where ϕ_{nc} is divided by λ. We argue that, similarly, U defines the scale for u_{nc} (for example see (3.16)). Hence, we argue that the two identities ϕ_{nc} and u_{nc} are *brought into context* through λ and U, respectively. We use this perceptual definition in order to propose procedures for determining the appropriate values for λ and U for an arbitrary problem class. We emphasize that in this process we utilize the multiple-cluster cost function carried in (3.12).

We suggest an imaginary situation, as depicted in Figure 3.1(a), in which two data items interact with two clusters ($N = 2$ and $C = 2$). The first data item, here x_1, is infinitely far from both clusters, in which case we expect $f_{11} = f_{12} = \frac{1}{2}$. The second data item, here x_2, however, is infinitely far from the second cluster and is at the distance of φ_f from the first one. Here, we ask what value of φ_f will result in $f_{21} = 2f_{22}$. This question can be asked in a different setting as follows: *For a data item which is infinitely far from two clusters, how close should it get to one cluster, while maintaining its distance to the other, in order for it to be favored by the former cluster two times than the latter?* One can show that setting $m = 2$ in the general case given in (3.12), we have [51],

$$f_{nc} = \frac{u_{nc}^{-1}}{\sum_{c'=1}^{C} u_{nc'}^{-1}}, \qquad (3.25)$$

based on which,

$$\lambda = \frac{\varphi_f}{u^{-1}\left(\frac{1}{2}\right)} = \varphi_f. \qquad (3.26)$$

In order to estimate the proper value for U, we utilize another imaginary situation in which one data item interacts with one cluster, as depicted in Figure 3.1(b). Here, we ask how far the data item should be from the cluster in order for it to be an outlier with a probability of half? This situation, in effect, defines the boundary of inliers and outliers when Maximum Likelihood is applied on p_n. We denote this distance as φ_p and utilize the fact that setting $m = 2$ in the general case given in (3.12) yields [51],

$$p_n = \frac{\sum_{c=1}^{C} u_{nc}^{-1}}{C\frac{1}{U} + \sum_{c=1}^{C} u_{nc}^{-1}}, \qquad (3.27)$$

and therefire,

$$U = u\left(\frac{1}{\lambda}\varphi_p\right). \qquad (3.28)$$

We note that φ_p and φ_f reflect different aspects of the relationship between data items and clusters. Nevertheless, conceptually, we can envision that one may want to set $\varphi_p = \varphi_f = \varphi_o$. In this line of reasoning, φ_o denotes the territory of a cluster, within which the cluster considers a data item as an inlier, therefore $p_x \geq \frac{1}{2}$, and also owns the data item when in competition with another farther cluster, therefore $f_{xc} \geq \frac{1}{2}$. Reworking (3.26) and (3.28) for $\varphi_p = \varphi_f = \varphi_o$ we arrive at $\lambda = \varphi_o$ and $U = \frac{1}{2}$.

Chapter 4

Experimental Results

The algorithm developed in this work is implemented as the Matlab class *Annina*. This class contains the core operations defined in this work, which, as stated, are class-independent. The child classes of this class override the two virtual functions $\phi(\cdot)$ and $\Psi(\cdot)$ and set the value of λ according to the specifications of their corresponding problem classes. These child classes also perform the load and visualization operations relevant to the cluster and data item models defined by the respective problem classes. In this work, we utilize six such child classes, as listed in Table 4.1.

The present work is, in effect, a search mechanism for discovering the local minimums of the single–cluster cost function associated with an arbitrary clustering problem. In this context, the method developed in this work utilizes a population of seed points on this multidimensional manifold and tallies the "distinct" cluster representations to whose "vicinity" a significant number of local searches lead. It also takes into consideration the relative number of times each of these "hyper-clusters" appear as the points of convergence.

The dimensionality of the search utilized in this work quickly grows out of the realm of perceptible geometries even for simple data item and cluster models. In effect, considering search spaces which can be visualizes in 3D, one is left with very few cases, one of which is carried in Figure 4.1.

Figure 4.1 shows the geometry of the search space for an arbitrary 2de problem instance. Here, each data item is a member of $\mathbb{R}^2$ and clusters, too, are points in $\mathbb{R}^2$. The manifold shown in Figure 4.1, clearly, contains three local minimums. Therefore, the purpose of the method developed in this work is, effectively, to "pour a large number of points on this manifold and to determine where these points end up and at what numbers". We now present the actual results of executing the proposed method on the input set of data items shown in Figure 4.1.

Table 4.1: Properties of the problem classes utilized in this paper.

Problem Class	x_n	ψ_c	Purpose	$\phi(x_n, \psi_c)$
2dc	$x_n \in \mathbb{R}^2$	$\psi_c = [m_c, \rho_c]$ $m_c \in \mathbb{R}^2$ $\rho_c > 0$	Finding Circles	$\left[\|x_n - m_c\|^2 - \rho_c^2\right]^2$
2de	$x_n \in \mathbb{R}^2$	$\psi_c \in \mathbb{R}^2$	Euclidean Clustering	$\|x_n - \psi_c\|^2$
2dl	$x_n \in \mathbb{R}^2$	$\psi_c = [m_c, v_c]$ $m_c \in \mathbb{R}^2$ $v_c \in \mathbb{R}^2, \|v_c\| = 1$	Finding Lines	$\|x_n - m_c - v_c^T(x_n - m_c)v_c\|^2$
3dpp	$x_n \in \mathbb{R}^3$	$\psi_c \in \mathbb{R}^3$	Finding Planes	$\frac{1}{\|\psi_c\|^2}\left(\psi_c^T x_n - \|\psi_c\|^2\right)^2$
ics	$x_n \in \mathbb{R}^3$	$\psi_c = [m_c, v_c]$ $m_c \in \mathbb{R}^3$ $v_c \in \mathbb{R}^3, \|v_c\| = 1$	Color Image Segmentation	$\|x_n - m_c - v_c^T(x_n - m_c)v_c\|^2$
ighe	$x_n \in \mathbb{R}$	$\psi_c \in \mathbb{R}$	Grayscale Image Segmentation	$(x_n - \psi_c)^2$

We point out that the input set of data items utilized in Figure 4.1 contains $C = 3$ clusters, wherein each cluster contains 50 data items and, that there are 50 additional outlier data items included in this set. We seed the developed method using $\hat{C} = 255$ *random* points in $\mathbb{R}^2$ and generate the DM shown in Figure 4.2(a). Note that this matrix is 255×255 and not 200×200. In other words, as discussed before, in the present work, in direct contrast with VAT and a significant majority of its variants, the size of the DM does not depend on N, the number of input data items.

Figure 4.2(a) shows the cluster DM corresponding to the input set of data items utilized in Figure 4.1. This matrix, natural to its definition, contains seemingly uncorrelated rows and columns. Nevertheless, through utilizing the conventional VAT reordering mechanism, the RDM shown in Figure 4.2(b) is generated. Note that this matrix demonstrates the typical characteristics of a "proper" the RDM, i.e. distinct dark blocks along the diagonal. Incidentally, and as expected, there are three dark blocks in Figure 4.2(b), which gracefully matches the fact that $C = 3$ clusters have been combined in order to generate the input set of data items utilized in this experiment.

As stated, the method developed in this work in fact does not utilize the RDM, and, more specifically, does not generate the RDM. In effect, the piece of information that the present method

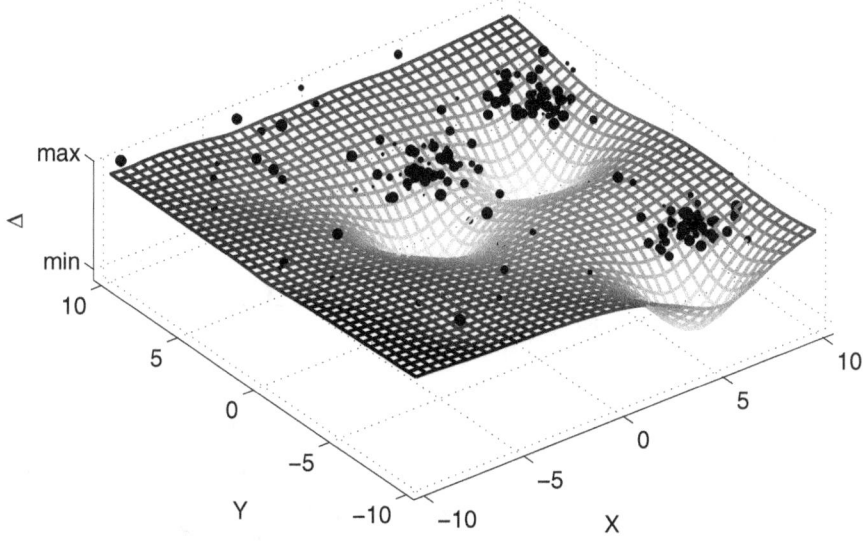

Figure 4.1: Geometry of the search space for an arbitrary 2de problem instance.

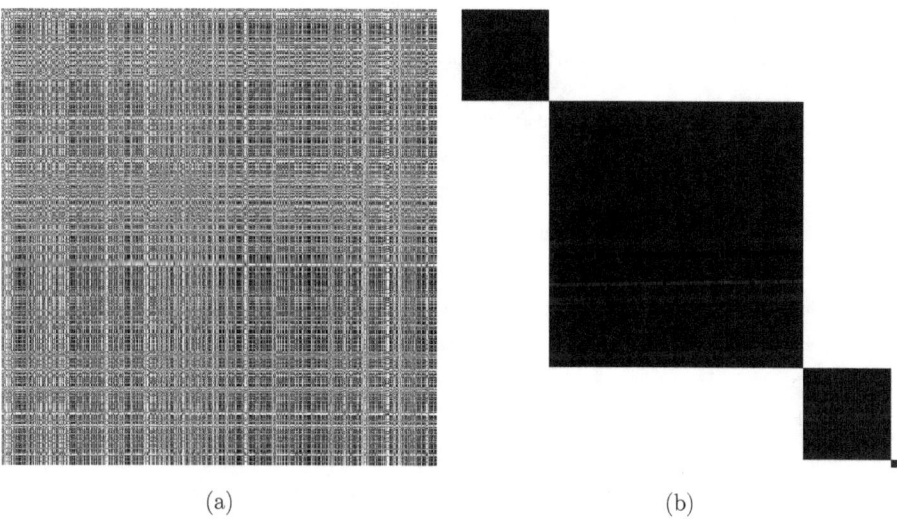

Figure 4.2: Cluster dissimilarity matrixes corresponding the input set of data items utilized in Figure 4.1. (a) DM. (b) RDM.

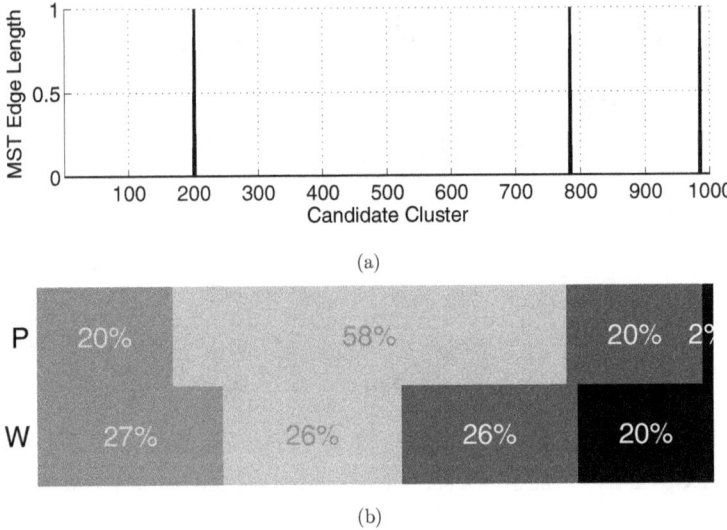

Figure 4.3: visualizations of the MST corresponding to the input set of data items utilized in Figure 4.1. (a) Successive MST edge lengths. (b) Prominence/Weight graph.

requires and generates is the MST. This information is carried in two different ways in Figure 4.3. Here, Figure 4.3(a) shows the length of the edges between the MST nodes as they are successively added to this tree. In this figure, the horizontal axis denotes the cluster seed index, thus ranging from 1 to $\hat{C}$, and the vertical axis denotes dissimilarity between pairs of clusters. Note that, in this representation, zero denotes two clusters which are absolutely identical, i.e. that the probability of mis-classification between them is zero. The value of one in this chart, on the other hand, denotes two completely complementary clusters, i.e. two cluster which do not own any mutual members or non-members.

The peaks in Figure 4.3(a) denote the points of separation between the sub-trees in the MST. In other words, each one of these peaks corresponds to a pair of clusters which are distinctly different. In effect, the discovery of these pairs is the main purpose behind seeding the single-cluster clustering framework developed in this work. Note that, although there are three peaks in Figure 4.3(a), and hence four subtrees in the corresponding MST, we only accept three clusters, because one is dropped due to the minimum prominence requirement. In other words, the short distance between the last peak in Figure 4.3(a) and the last data item denotes a local minimum which is not "prominent enough" in order to be accepted as a "stable" cluster. This is also visible in Figure 4.3(b), which denotes that about 2% of the cluster seeds fall into the subtree which corresponds to the gap between the last peak of the MST edge length graph and the last data item.

The visualization carried in Figure 4.3(b) displays cluster weight and prominence values. Note that, these two metrics are distinctly different in terms of what they measure. Prominence of a cluster, as stated before, denotes the ratio of the cluster seeds which fall into the vicinity of a convergence points in the cluster space. Weight of a cluster, on the other hand, denotes the ratio of the population of the input set of data items which are classified into a cluster. Nevertheless, weight and prominence, both, add up to one for any set of clusters. We note that the top row in Figure 4.3(b) denotes cluster prominence, hence the indicator "P", and the bottom row denotes cluster weight, hence the indicator "W".

Assessment of cluster weights in Figure 4.3(b) shows that about a quarter of the data items are not classified in either of the three *prominent* subtrees. This number beautifully matches $\frac{50}{3 \times 50 + 50}$. In other words, in the input set of data items utilized in this experiment, there are 50 outlier data items and three sets of data items, each with 50 data items, corresponding to each cluster. The fact that the three clusters generated by the developed method each contains about a quarter of the data items and a quarter of the data items is left as outlier is worth extra attention, and appreciation, in this experiment. Cluster prominent, too, paints an important picture. Through analyzing the cluster prominent figures carried in Figure 4.3(b) we recognize that about 60% of the cluster seeds fall into one special cluster. Cross-examination of this number with Figure 4.4, which carries the visualization of the set of clusters produced in this experiment, is informative.

Figure 4.4 shows that the developed algorithm finds three clusters in the input set of data items. Note that, the shaded circles drawn in Figure 4.4 denote one of the cluster seeds which corresponds to any of the final clusters. Through the cross-examination of Figures 4.3(b) and 4.4, we recognize that the cluster which is close to the origin is in fact the "most prominent" one. This cluster, as seen in Figure 4.1, corresponds to a large "basin" which "absorbs" a majority of the cluster seeds which happen to fall in its vicinity.

Figure 4.5 shows the paths of the cluster seeds towards their point of convergence in this experiment. We emphasize that this visualization is in fact possible and meaningful specifically and explicitly because data item and cluster models are identical in this particular problem class. In other words, in this particular problem class, one can meaningfully present a data item alongside

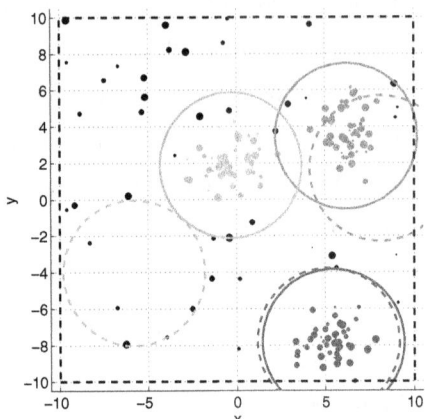

Figure 4.4: Set of clusters produced by the developed method for the data shown in Figure 4.1.

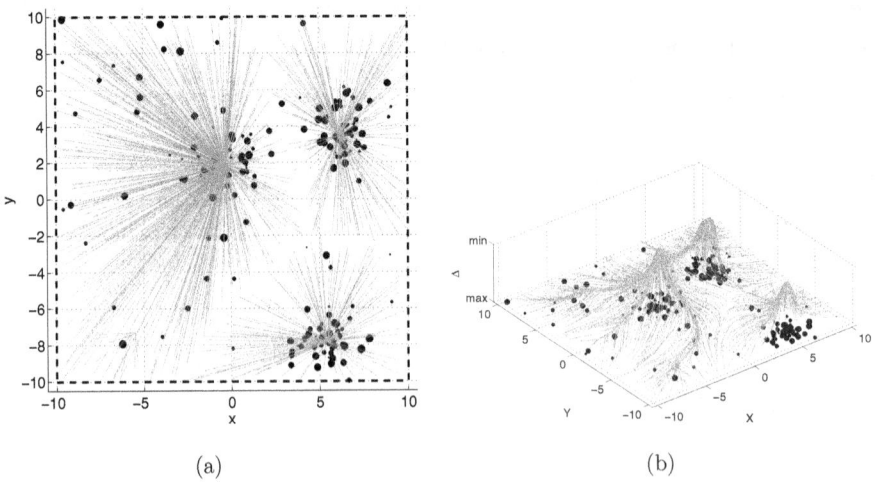

Figure 4.5: Convergence paths of the cluster seeds corresponding to the clusters shown in Figure 4.4. (a) Two dimensional view from top. (b) Three dimensional view of cluster representations and the associated cost function values.

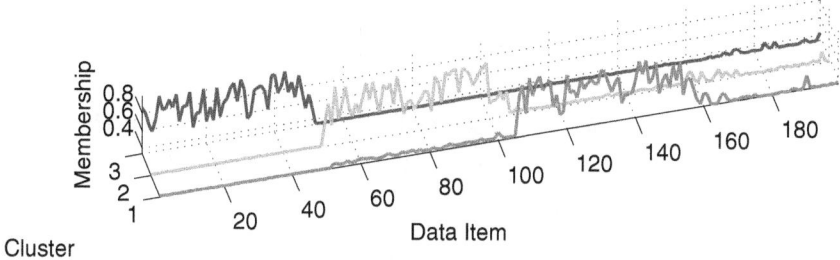

Figure 4.6: Membership levels of the reordered input data items to the converged clusters shown in Figure 4.4.

a cluster, and that is exactly what has made Figure 4.5 possible. It is also important to note that Figures 4.5(a) and (b) show the same data from two different vantage points. In fact, Figure 4.5(a) merely shows the line which connects a cluster seed to its converged status. Hence, the number of lines which converge to the middle cluster in Figure 4.5(a) and the large prominence of this cluster, as seen in Figure 4.3(b), are two different ways to carry the information that this particular cluster is positioned so that many randomly positioned clusters converge to it. One may in fact directly count the number of lines which converge to each final cluster in Figure 4.5(a) and generate the prominence graph carried in Figure 4.3(b).

The visualization shown in Figure 4.5(b) is another, more detailed, representation of the data carried in Figure 4.5(a). Here, we have included the intermediate positions of each cluster seed as well. Additionally, we have converted $\psi \in \mathbb{R}^2$ to $[\psi, \Delta] \in \mathbb{R}^3$. Note that the direction of the z-axis in Figure 4.5(b) is inverted. Hence, an optimization problem which seeks local minimums has generated (upside-down) hills. Nevertheless, it is informative to observe the lines, each of which corresponds to one clusters, as they merge together and form the peaks which correspond to the cluster representations generated by the developed method.

Moreover, we emphasize the existence of the disturbance present at the bottom left of Figure 4.5(a), and less clearly visible at the same position in Figure 4.5(b). This small group of converging cluster paths in fact corresponds to the two adjacent peaks in Figure 4.3(a) and the fourth non-prominent cluster in Figure 4.3(b). In other words, a small minority of cluster seeds do end up converging to the same cluster representation, but due to the fact that the prominence of this group is negligible, their mutual agreement is (correctly) ignored by the developed algorithm.

Figure 4.6 shows the results of reordering the input data items according to their level of membership to the clusters discovered by the proposed method. Here, the reordering process utilizes the following *index*,

$$i_n = \sum_{c=1}^{C} 2^{c-1} \left[p_{nc} > \frac{1}{2} \right],\qquad(4.1)$$

We note that this membership output has similarities to the profile graphs generated by reVAT [39]. Nevertheless, these graphs are generated by the developed method *after* the MST is dissected and *after* the prominent clusters have been selected. reVAT, on the other hand, performs thresholding on the DM in order to generate the membership graphs. Additionally, in direct contrast to reVAT, membership graphs are mere byproducts of the process employed by the developed algorithm and are in no way instrumental in the operation.

As stated, this work utilizes a Bayesian measure of cluster dissimilarity. Figure 4.7 exhibits one case of executing this measure. Here, Figures 4.7(a) and (b) show cluster candidates #12 and #21 for an arbitrary 2de problem instance. In this case, $\delta_{12,21}$ is calculated as 0.92. Figure 4.7(c) shows the values of p_n corresponding to these two clusters. Note that the data items are reordered in this visualization in order to enhance the spatial contiguity of the membership graphs. We observe in this experiment that the employed cluster dissimilarity measure in fact *correctly* determines that these two clusters are *dissimilar*. In other cases, wherein the pair of clusters are similar, we observe that the dissimilarity is *correctly* assessed to be close to zero as well.

The proposed algorithm contains a local search and aggregation procedure which is stochastic in its essence. Hence, not only there is no guarantee for the repeatability of the outcome of the algorithm for any particular problem instance, but also, and more importantly, there is no way to predict what the algorithm will produce for a problem instance. Nevertheless, our experiments indicate that there is a *high degree* of repeatability for the output of the developed algorithm. Here, we present the results of five independent executions of the proposed algorithm on an arbitrary 2dl problem instance.

Figure 4.8(a) shows a 2dl problem instance and Figures 4.8(b) to (e) exhibit the results of five independent executions of the developed algorithm. In each of these five figures, the solid lines indicate individual clusters and the data items are colored according to the cluster they are classified to. The dashed lines in these figures demonstrate the cluster seeds which have converged to the corresponding cluster. Hence, we observe that, one, in all cases, the developed algorithm converges to three clusters. It is important to emphasize that the developed algorithm does not have *a priori*

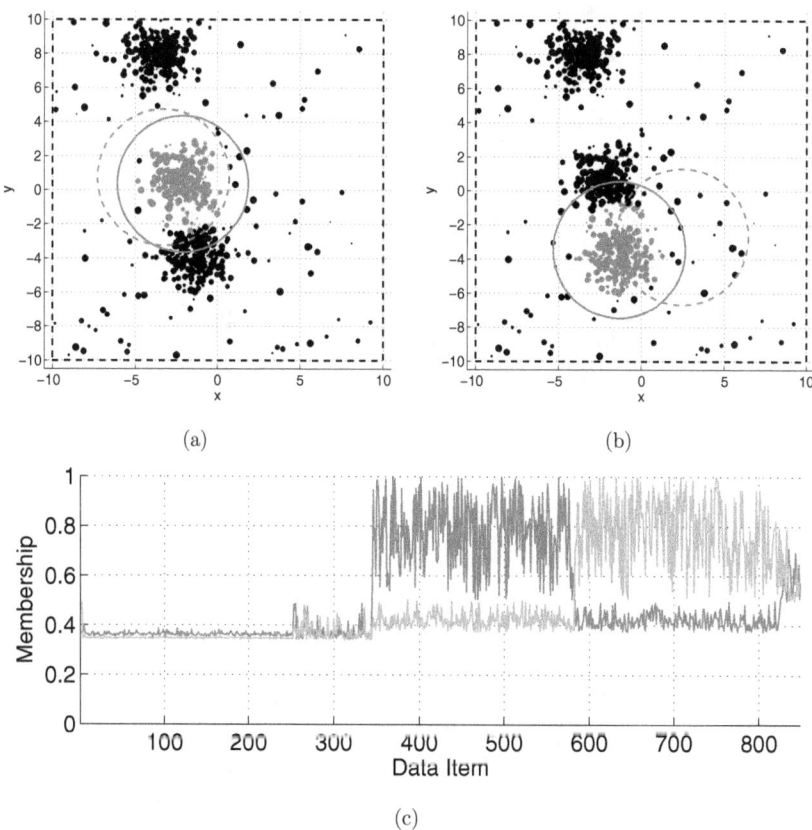

Figure 4.7: Assessment of the cluster dissimilarity measure utilized in this work for an arbitrary 2de problem instance. (a) and (b) Two cluster candidates produced by the developed algorithm. (c) Reordered membership graphs for the two clusters carried in (a) and (b).

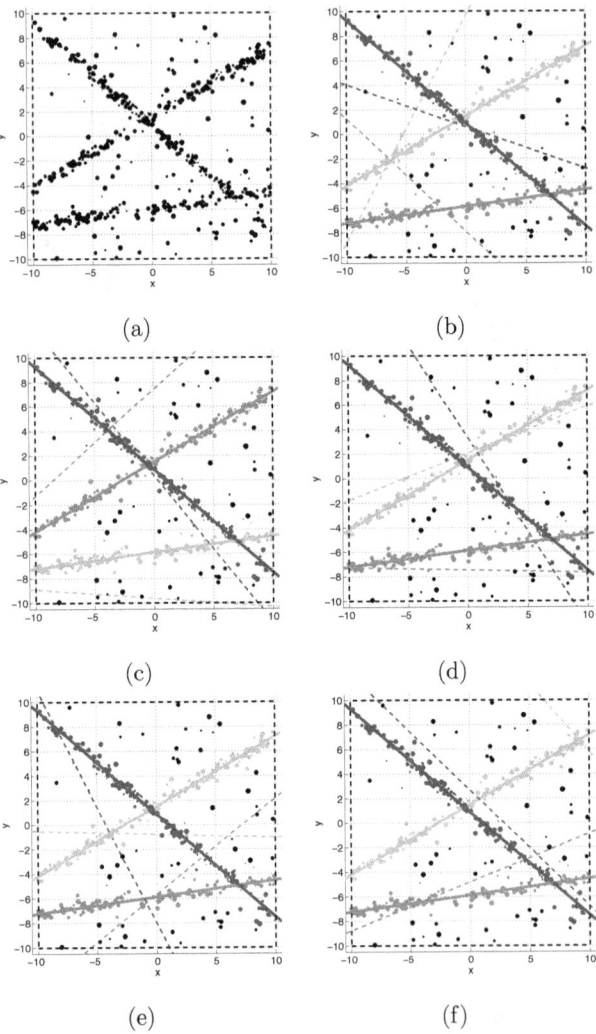

Figure 4.8: Repeatability assessment for the 2dl problem instance shown in (a). (b), (c), (d), (e) and (f) exhibit the results of five independent executions of the developed algorithm.

access to information about the number of clusters which it is supposed to look for. Additionally, and equally importantly, the clusters generated during these five independent executions appear to be identical. We in fact performed analytical comparison on the classification results corresponding to these five runs and calculated the pairwise similarity matrix and found it to be $\mathbf{1}_{5\times 5}$, within the numerical accuracy of the calculations. In other words, these five runs produce the same *exact* classification results. We emphasize that this situation is exceptional and we expect the similarity matrix to contain more realistic values, which are still close to one, wherein some disturbance is caused by noise and other factors. Note that the order of the clusters does vary amongst the results carried in Figures 4.8(b) to (e). In other words, we observe that the developed algorithm can be expected to produce the same clusters, potentially permuted differently, in independent executions. In this experiment, $\mathbf{X}$ contains 528 data items and we utilized $\hat{C} = 255$. The developed algorithm converged to 3 clusters in all cases. The five executions of the developed algorithm in this experiment took $5,334 \pm 197$ milliseconds to converge.

In the present work, $\hat{C}$ governs the processing budget of the algorithm. In other words, $\hat{C}$ maintains the number of independent attempts to find local minimums in the optimization manifold. Hence, one would expect to receive more representative clusters as $\hat{C}$ grows. Additionally, larger $\hat{C}$ values enhance the accuracy of the estimation of cluster prominence values. Hence, larger $\hat{C}$ is expected to lead to, both, the discovery of more useful clusters as well as the more confident determination that they are indeed more stable clusters. Nevertheless, a larger $\hat{C}$ value directly translates into a more expensive algorithm. This double-sided relationship between the value of $\hat{C}$ and the *quality* of the outputs of the developed algorithm is the topic of the next experiment.

Figure 4.9 shows the impact of the value of $\hat{C}$ on the converged clusters for a *3dpp* problem instance. This problem class is concerned with finding planar sections in range data captured by a *Kinect 2* sensor. The depth-maps used in this experiment are captured at the resolution of 424×512 pixels. Here, intrinsic parameters of the camera are acquired through the Kinect SDK and each data item in this problem class has the weight of one. The data utilized in this experiment corresponds to a corner in a room, wherein a human body is present. Hence, we expect to receive three clusters, two of which ought to identify the walls in the room and the third should correspond to the floor.

Here, we use the following values for $\hat{C}$: 5, 10, 20, 40, 80, and 100, and observe, as seen in Figures 4.9(a) and (b), that for small $\hat{C}$, the generated clusters correspond to local minimums in the search space. As $\hat{C}$ increases to 40 and 80 we notice that the proposed algorithm converges to

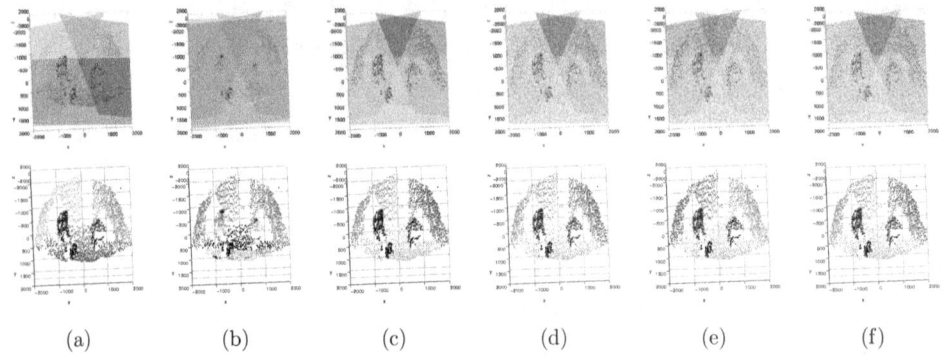

Figure 4.9: Impact of $\hat{C}$ on the outputs of the developed algorithm for a 3dpp problem instance. Top: Converged clusters. Bottom: Classification results. (a) $\hat{C} = 5$. (b) $\hat{C} = 10$. (c) $\hat{C} = 20$. (d) $\hat{C} = 40$. (e) $\hat{C} = 80$. (f) $\hat{C} = 100$. Input data courtesy of *Epson Edge, Epson Canada Limited*.

the three desired clusters. Hence, we observe that the behavior of the developed method follows common logic in this respect; i.e. a certain minimum processing budget is required for the search to be encompassing and solid, and, additionally and importantly, investing more processing power does not either increase or decrease the quality of the outputs.

We note, however, that the order of complexity of the developed algorithm contains both a linear as well as a quadratic function of $\hat{C}$. In order to assess this situation, we review the time elapsed algorithm by the developed as it completes the clustering process. As seen in Figure 4.10(b) the execution time of the developed algorithm increases linearly with $\hat{C}$. We note that the linear inflation of the time taken by the developed algorithm in order to converge, relates to the time required for the processing of the $\hat{C}$ cluster seeds. The quadratic component, however, corresponds to the generation of the MST, and as seen in Figure 4.10(b) the latter appears to be less significant than the former.

Finally, Figures 4.11, 4.12, and 4.13 carry sample results generated for problem instances corresponding to the six problem classes carried in Table 4.1. In these figures, every row corresponds to one problem instance. The columns of every row contain the representations of different aspects of the corresponding problem instance as it goes through the proposed algorithm. There, in every case, the first column carries the input set of data items. The second column in these figures contains the RDM, and the last column exhibits the results of the classification of the input set of

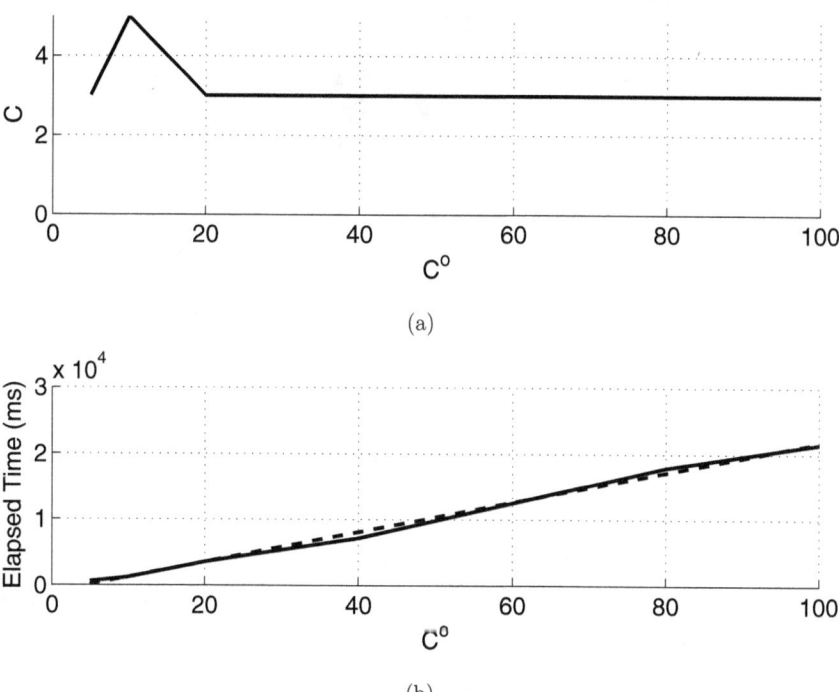

Figure 4.10: Number of discovered clusters and the elapsed time for the experiments shown in Figure 4.9. (a) Number of clusters. (b) Elapsed time.

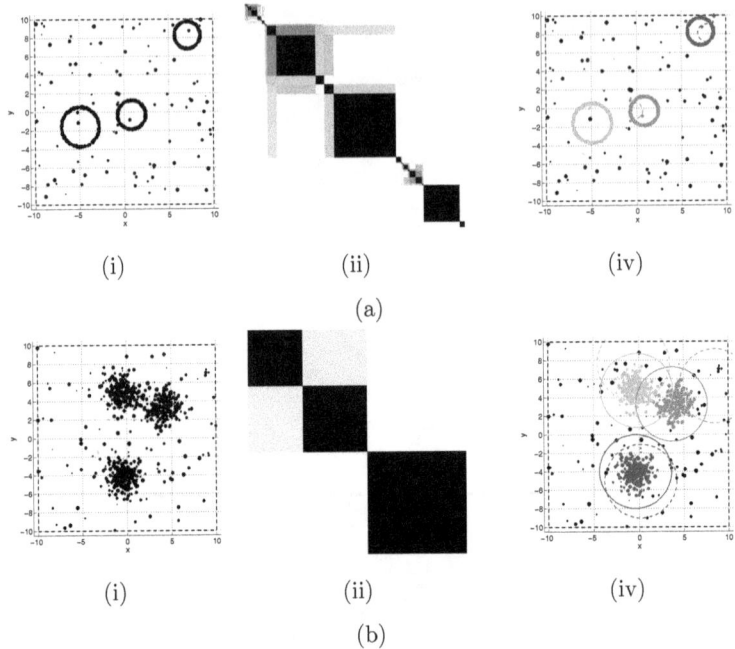

Figure 4.11: Sample results generated by the proposed algorithm for problem instances corresponding to the six problem classes carried in Table 4.1. (a) 2dc. (b) 2de.

data items according to the clusters generated by the proposed algorithm. The third column, if present, exhibits the representation of the clusters generated by the developed algorithm.

Figure 4.11(a) presents a 2dc problem instance. This problem class is concerned with identifying circles in a weighted set of 2D points. In this problem instance, as seen in Figure 4.11(a)(i), there are three circles, and, as seen in Figure 4.11(a)(iv), the developed algorithm correctly identifies these clusters. Figure 4.11(a)(ii) shows the RDM corresponding to this problem instance. Here, the three dark diagonal blocks correspond to the three clusters generated by the developed algorithm. The input set of data items utilized in this experiment contains 850 data items and the proposed algorithm takes 6,688 milliseconds to converge for this problem instance.

Similarly, Figures 4.11(b) and 4.12(a) present 2de and 2dl problem instances, respectively. The former problem class corresponds to finding Euclidean compact sets and the latter problem class involves finding linear sets, both among weighted points on a 2D plane. The problem instance depicted in Figure 4.11(b) contains 850 input data items and the developed algorithm requires 5,391

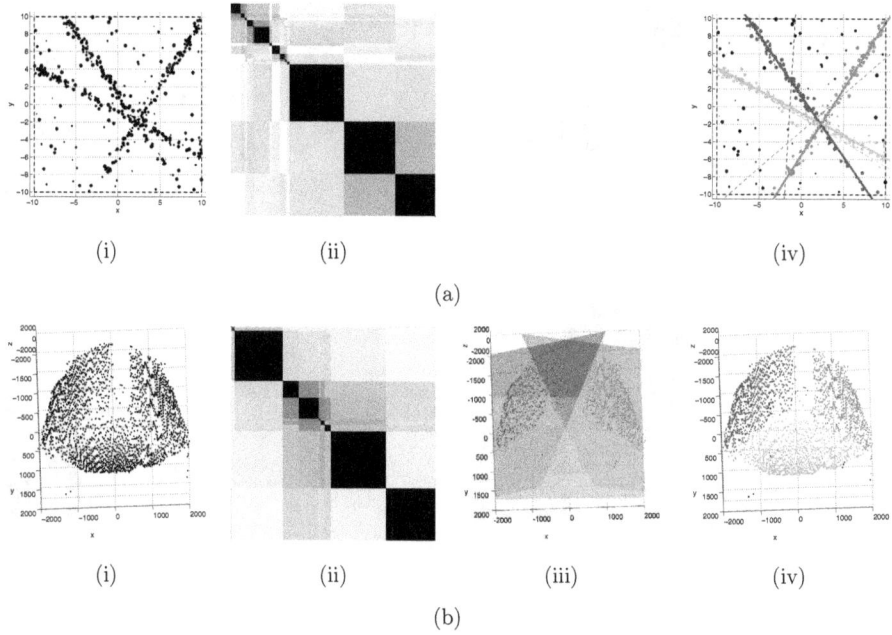

Figure 4.12: Sample results generated by the proposed algorithm for problem instances corresponding to the six problem classes carried in Table 4.1. (a) 2dl. (b) 3dpp. Input data courtesy of *Epson Edge, Epson Canada Limited.*

milliseconds to converge for this problem instance. The problem instance carried in Figure 4.12(a) contains 483 input data items and the developed algorithm takes 4,750 milliseconds to converge for it.

In contrast to Figures 4.11(a), fig:samples:one(b), and fig:samples.two(a), which utilize artificially generated input data items, the next three problem instances, i.e. the ones carried in Figures 4.12(b), fig:samples:three(a), and fig:samples:three(b), utilize input data items which have in fact been *measured*. These problem classes include 3dpp, i.e. finding planes in a 3D space, ics, i.e. segmentation of a color image using a linear model of homogeneity [73, 74], and ighe, i.e. segmentation of a grayscale image using its histogram, respectively.

The problem instance carried in Figure 4.12(b)(i) contains 2,759 input data items and the result shown in Figure 4.12(b)(iv) is generated in 62,031 milliseconds. Similarly, the input set of data items utilized for the experiment carried in Figure 4.13(a) contains 16,384 input data items

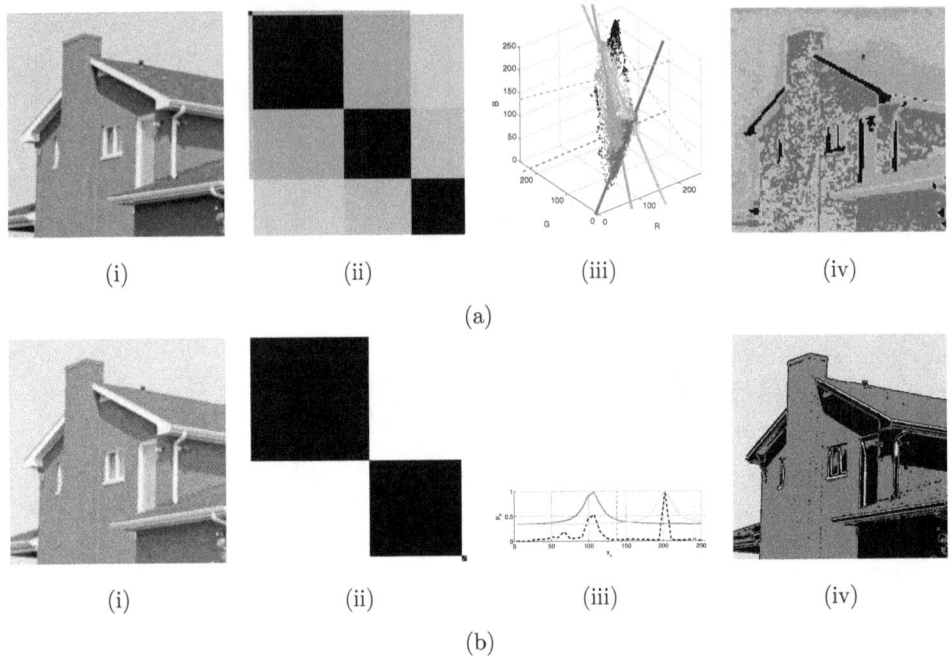

Figure 4.13: Sample results generated by the proposed algorithm for problem instances corresponding to the six problem classes carried in Table 4.1. (a) ics. (b) ighe.

and the developed algorithm converges for this problem instance in 42,531 milliseconds. Finally, the problem instance corresponding to Figure 4.13(b)(i) contains 32 bins and the result carried in Figure 4.13(b)(iv) is generated in 3,625 milliseconds.

Nevertheless, there is no implied or assumed claim in this manuscript that the present algorithm functions *desirably* for *every* problem instance. Figure 4.14 carries two *failed* executions corresponding to 2dc and 2de problem instance, wherein, either an undesired cluster is "discovered" in the data or that a desired cluster is ignored.

In Figure 4.14(a) we observe that, although only two apparent clusters exist in Figure 4.14(a)(i), but the corresponding RDM, shown in Figure 4.14(a)(ii), contains other dark blocks, which translate into the undesired third cluster shown in Figure 4.14(a)(iii). This problem instance contains 742 data items and the developed algorithm converges for it in 4,766 milliseconds. The opposite situation occurs for the problem instance shown in Figure 4.14(b), in which, although there exists a dark block corresponding to the third cluster in the input set of data items, as seen in Fig-

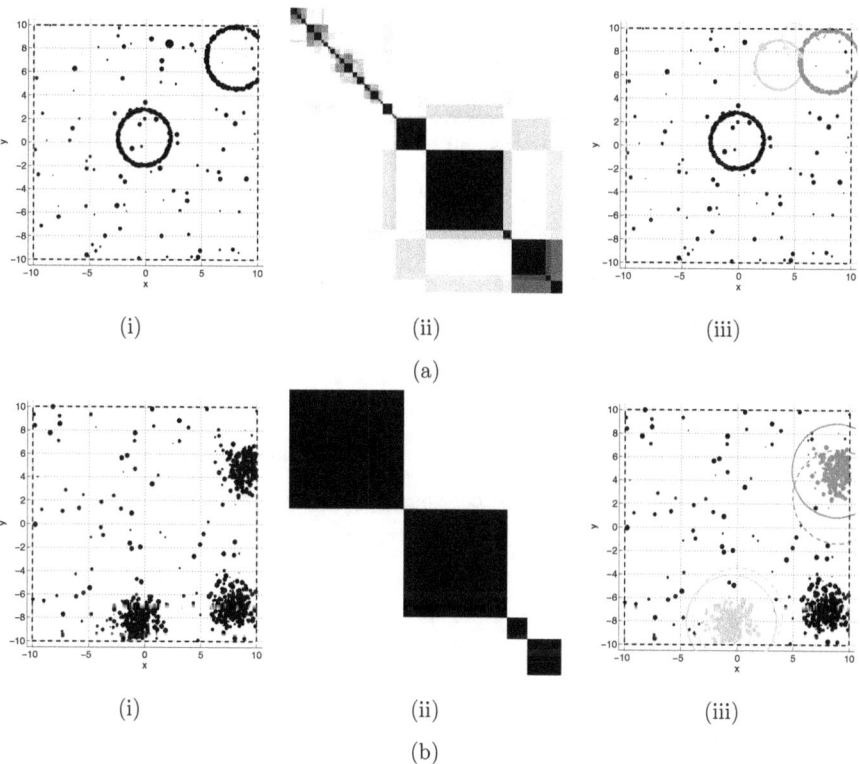

Figure 4.14: Undesired results generated by the proposed algorithm. (a) 2dc. (b) 2dc.

ure 4.14(a)(ii), but this block is not prominent enough, and is therefore ignored, thus resulting in one desired cluster not identified, as seen in Figure 4.14(a)(iii). The next step of this work is to assess these and other suboptimal situations and to devise mitigation strategies for this deficiency.

As discussed, a majority of other algorithms in the field utilize the DM corresponding to the input set of data items. In addition to this notion being irrelevant for many models of homogeneity, the computational complexity of that process is of $O(N^2)$. The present work, however, generates the DM at the level of clusters, and hence its computational complexity is of $O(\hat{C}^2)$. In fact, not only $\hat{C} \ll N$, but also $\hat{C}$ relates to the number of cluster in the input set of data items and not the number of data items, and, hence, the developed algorithm is *categorically* faster than the methods available in the literature. Nevertheless, as the complexity of the homogeneity models grows, $\hat{C}^2$ indeed becomes a challenge. Hence, in the continuation of this work, we intend to address this aspect of the present work as well.

Chapter 5

Conclusions

The present work in no way claims to provide a final solution to the generic problem of unsupervised clustering in a class-independent context. What this work aims to achieve is, first, to utilize VAT-style dissimilarity assessment and reordering at the level of clusters. This is in direct contrast to a majority of the works in the literature, which generate the dissimilarity matrix for the input data items. That paradigm is not only inefficient, but also, and more importantly, it is based on the implicit assumption of prototype-based clustering. This is essentially because available VAT-style algorithms utilize the direct Euclidean distance between data items, whilst this entity is in effect irrelevant to the clusterability of the data in any context except for relational clustering. Moreover, the inefficiency of classical VAT algorithms is due to the fact that their computational complexity is of $O(N^2)$, where N is the number of data items. As such, we demonstrate in this work that the VAT reordering mechanism can be utilized at the level of clusters. This is important not only because the computational complexity of the resulting algorithm will be of $O(N)$, but also, and more importantly, because in this proposed framework one can discuss clustering in the context of general notions of homogeneity. Nevertheless, the present work does not guarantee the usability of the outcomes of the developed process in every execution for any problem instance. Moreover, the computational complexity of the developed algorithm, while down from quadratic to linear, is still a prohibiting factor in some caes. We intend to focus on these aspects of the present work in the continuation of this research.

Acknowledgments

We wish to thank the management of *Fio Corporation* for their support. We thank the management of *Epson Edge, Epson Canada Limited*, for allowing us to use the Kinect data utilized in some of the experiments carried in this paper. The author wishes to thank *Prof. James C. Bezdek* for his mentorship and for providing us with critical pieces of the literature. The majority of this research was carried out while listening to the music posted on the SoundCloud channel maintained by *Sahar Samimi*. The author wishes to thank her. The author would like to thank *Sahar Bastanirad* for proofreading this manuscript.

Bibliography

[1] G. Guest, E. McLellan, Distinguishing the trees from the forest: Applying cluster analysis to thematic qualitative data, Field Methods 15 (2003) 186–201.

[2] J. C. Bezdek, R. J. Hathaway, VAT: A tool for visual assessment of (cluster) tendency, in: Proceedings of the 2002 International Joint Conference on Neural Networks (IJCNN 2002), Vol. 3, 2002, pp. 2225–2230.

[3] J. C. Dunn, Indices of partition fuzziness and the detection of clusters in large data sets, in: B. G. M.M. Gupta, G.N. Saridis (Ed.), Automata and Decision Process, North-Holland Publishers, Amsterdam, 1977, pp. 271–283.

[4] L. Hubert, P. Arabie, Comparing partitions, Journal of Classification 2 (1) (1985) 193–218.

[5] J. C. Dunn, A fuzzy relative of the ISODATA process and its use in detecting compact well-separated clusters, Journal of Cybernetics 3 (3) (1973) 32–57.

[6] D. L. Davies, D. W. Bouldin, A cluster separation measure, IEEE Transactions on Pattern Analysis and Machine Intelligence PAMI-1 (2) (1979) 224–227.

[7] M. K. Pakhira, S. Bandyopadhyay, U. Maulik, Validity index for crisp and fuzzy clusters, Pattern Recognition 37 (3) (2004) 487–501.

[8] J. C. Bezdek, Q. W. Li, Y. Attikiouzel, M. Windham, A geometric approach to cluster validity for normal mixtures, Soft Computing 1 (4) (1997) 166–179.

[9] G. W. Milligan, M. C. Cooper, An examination of procedures for determining the number of clusters in a data set, Psychometrika 50 (2) (1985) 159–179.

[10] R. J. Hathaway, J. C. Bezdek, Visual cluster validity for prototype generator clustering models, Pattern Recognition Letters 24 (9–10) (2003) 1563–1569.

[11] L. Freeman, Displaying hierarchical clusters, Connections 17 (2) (1994) 46–52.

[12] S. Gaddam, Clustertendency repository on github, https://github.com/venganesh/ClusterTendency (Accessed in December 2015).

[13] R. J. Hathaway, J. C. Bezdek, Extending fuzzy and probabilistic clustering to very large data sets, Computational Statistics & Data Analysis 51 (1) (2006) 215–234.

[14] R. C. Prim, Shortest connection networks and some generalizations, Bell System Technical Journal 36 (1957) 1389–1401.

[15] K. Rosen, Discrete Mathematics and Its Applications, McGraw-Hill, New York, 1999.

[16] P. Prabhu, K. Duraiswamy, Enhanced VAT for cluster quality assessment in unlabeled datasets, Journal of Circuits, Systems and Computers 21 (01) (2012) 1–19.

[17] N. Otsu, A threshold selection method from gray level histograms, IEEE Transactions on Systems, Man, and Cybernetics 9 (1) (1979) 62–66.

[18] R. L. Ling, A computer generated aid for cluster analysis, Communications of the ACM 16 (6) (1973) 355–361.

[19] R. Johnson, D. W. Wichern, Applied Multivariate Statistical Analysis, Prentice-Hall, Englewood Cliffs, 1992.

[20] R. Sibson, SLINK: An optimally efficient algorithm for the single-link cluster method, The Computer Journal 16 (1) (1973) 30–34.

[21] M. S. Aldenderfer, R. K. Blashfield, Computer programs for performing hierarchical cluster analysis, Applied Psychological Measurement 2 (3) (1978) 403–411.

[22] T. C. Havens, J. C. Bezdek, J. M. Keller, M. Popescu, J. M. Huband, Is VAT really single linkage in disguise?, Annals of Mathematics and Artificial Intelligence 55 (3) (2009) 237–251.

[23] D. Kumar, J. C. Bezdek, M. Palaniswami, S. Rajasegarar, C. Leckie, T. C. Havens, A hybrid approach to clustering in big data, IEEE Transactions on Cybernetics PP (99) (2015) 1–1. doi:10.1109/TCYB.2015.2477416.

[24] J. M. Huband, J. C. Bezdek, R. J. Hathaway, bigVAT: Visual assessment of cluster tendency for large data sets, Pattern Recognition 38 (11) (2005) 1875–1886.

[25] L. Wilkinson, M. Friendly, The history of the cluster heat map, The American Statistician 63 (2) (2009) 179–184.

[26] T. Loua, Atlas statistique de la population de Paris, Dejey, Paris, 1873.

[27] W. C. Brinton, Graphic Methods for Presenting Facts, The Engineering Magazine Company, New York, 1914.

[28] J. N. Weinstein, A postgenomic visual icon, Science 319 (5871) (2008) 1772–1773.

[29] M. Hahsler, K. Hornik, C. Buchta, Getting things in order: An introduction to the R package seriation, Journal of Statistical Software 25 (1).

[30] M. Hahsler, K. Hornik, Dissimilarity plots: A visual exploration tool for partitional clustering, Journal of Computational and Graphical Statistics 20 (2) (2011) 335–354.

[31] J. C. Bezdek, R. J. Hathaway, J. M. Huband, Visual assessment of clustering tendency for rectangular dissimilarity matrices, IEEE Transactions on Fuzzy Systems 15 (5) (2007) 890–903.

[32] L. A. F. Park, J. C. Bezdek, C. A. Leckie, Visualization of clusters in very large rectangular dissimilarity data, in: Proceedings of 4th International Conference on Autonomous Robots and Agents (ICARA 2009), 2009, pp. 251–256.

[33] T. C. Havens, J. C. Bezdek, A new formulation of the coVAT algorithm for visual assessment of clustering tendency in rectangular data, International Journal of Intelligent Systems 27 (6) (2012) 590–612.

[34] T. C. Havens, J. C. Bezdek, An efficient formulation of the improved visual assessment of cluster tendency (iVAT) algorithm, IEEE Transactions on Knowledge and Data Engineering 24 (5) (2012) 813–822.

[35] R. J. Hathaway, J. C. Bezdek, J. M. Huband, Scalable visual assessment of cluster tendency for large data sets, Pattern Recognition 39 (7) (2006) 1315–1324.

[36] L. Wang, U. T. Nguyen, J. C. Bezdek, C. A. Leckie, K. Ramamohanarao, iVAT and aVAT: Enhanced visual analysis for cluster tendency assessment, in: M. J. Zaki, J. X. Yu, B. Ravindran, V. Pudi (Eds.), Advances in Knowledge Discovery and Data Mining, Vol. 6118 of Lecture Notes in Computer Science, Springer Berlin Heidelberg, 2010, pp. 16–27.

[37] P. Prabhu, K. Duraiswamy, An efficient visual analysis method for cluster tendency evaluation, data partitioning and internal cluster validation, Computing and Informatics 32 (5) (2013) 1013–1037.

[38] J. M. Huband, J. C. Bezdek, Computational intelligence: Research frontiers, Springer Berlin Heidelberg, Berlin, Heidelberg, 2008, Ch. VCV2 - Visual Cluster Validity, pp. 293–308.

[39] J. M. Huband, J. C. Bezdek, R. J. Hathaway, Revised visual assessment of (cluster) tendency (reVAT), in: Proceedings of IEEE Annual Meeting of the Fuzzy Information (NAFIPS 2004), Vol. 1, 2004, pp. 101–104.

[40] T. C. Havens, J. C. Bezdek, M. Palaniswami, Scalable single linkage hierarchical clustering for big data, in: Proceedings of 2013 IEEE Eighth International Conference on Intelligent Sensors, Sensor Networks and Information Processing, 2013, pp. 396–401.

[41] L. Wang, X. Geng, J. C. Bezdek, C. Leckie, R. Kotagiri, SpecVAT: Enhanced visual cluster analysis, in: Proceedings of 2008 Eighth IEEE International Conference on Data Mining, 2008, pp. 638–647.

[42] I. J. Sledge, T. C. Havens, J. M. Huband, J. C. Bezdek, J. M. Keller, Finding the number of clusters in ordered dissimilarities, Soft Computing 13 (12) (2009) 1125–1142.

[43] L. Wang, C. Leckie, K. Ramamohanarao, J. C. Bezdek, Automatically determining the number of clusters in unlabeled data sets, IEEE Transactions on Knowledge and Data Engineering 21 (3) (2009) 335–350.

[44] T. C. Havens, J. C. Bezdek, J. M. Keller, M. Popescu, Dunn's cluster validity index as a contrast measure of VAT images, in: Proceedings of the 19th International Conference on Pattern Recognition (ICPR 2008), 2008, pp. 1–4.

[45] T. C. Havens, J. C. Bezdek, J. M. Keller, M. Popescu, Clustering in ordered dissimilarity data, International Journal of Intelligent Systems 24 (5) (2009) 504–528.

[46] V. Ganti, J. Gehrke, R. Ramakrishnan, Mining very large databases, Computer 32 (8) (1999) 38–45.

[47] Y. Ding, R. F. Harrison, Relational visual cluster validity (RVCV), Pattern Recognition Letters 28 (15) (2007) 2071–2079.

[48] M. Popescu, J. C. Bezdek, J. M. Keller, T. C. Havens, J. M. Huband, A new cluster validity measure for bioinformatics relational datasets, in: Proceedings of IEEE World Congress on Computational Intelligence Fuzzy Systems (FUZZ-IEEE 2008), 2008, pp. 726–731.

[49] D. Kumar, J. C. Bezdek, S. Rajasegarar, C. Leckie, M. Palaniswami, A visual-numeric approach to clustering and anomaly detection for trajectory data, The Visual Computer (2015) 1–17.

[50] J. C. Bezdek, Pattern Recognition with Fuzzy Objective Function Algorithms, Plenum Press, New York, 1981.

[51] A. Abadpour, Rederivation of the fuzzypossibilistic clustering objective function through Bayesian inference, Fuzzy Sets and Systems 305 (2016) 29–53.

[52] P. D'Urso, Fuzzy clustering of fuzzy data, in: J. V. de Oliveira, W. Pedrycz (Eds.), Advances in Fuzzy Clustering and its Applications, Wiley, England, 2007, pp. 155–192.

[53] A. Abadpour, A. S. Alfa, J. Diamond, Video-on-demand network design and maintenance using fuzzy optimization, IEEE Transactions on Systems, Man, and Cybernetics, Part B: Cybernetics 38 (2) (2008) 404–420.

[54] R. J. Hathaway, Y. Hu, Density-weighted fuzzy C-means clustering, IEEE Transactions on Fuzzy Systems 17 (1) (2009) 243–252.

[55] R. Krishnapuram, J. M. Keller, A possibilistic approach to clustering, IEEE Transactions on Fuzzy Systems 1 (2) (1993) 98–110.

[56] G. Beni, X. Liu, A least biased fuzzy clustering method, IEEE Transactions on Pattern Analysis and Machine Intelligence 16 (9) (1994) 954–960.

[57] R. N. Dave, Characterization and detection of noise in clustering, Pattern Recognition Letters 12 (11) (1991) 657–664.

[58] R. N. Dave, Robust fuzzy clustering algorithms, in: Second IEEE International Conference on Fuzzy Systems, Vol. 2, 1993, pp. 1281–1286.

[59] R. N. Dave, R. Krishnapuram, Robust clustering methods: A unified view, IEEE Transactions on Fuzzy Systems 5 (2) (1997) 270–293.

[60] R. Kruse, C. Doring, M.-J. Lesot, Fundamentals of fuzzy clustering, in: J. V. de Oliveira, W. Pedrycz (Eds.), Advances in Fuzzy Clustering and its Applications, Wiley, England, 2007, pp. 3–29.

[61] P. W. Holland, R. E. Welsch, Robust regression using iteratively reweighted least squares, Communication Statistics - Theory and Methods A6 (9) (1977) 813–827.

[62] G. Wesolowski, The Weber problem: History and perspective, Location Science 1 (1993) 5–23.

[63] A. Abadpour, A sequential bayesian alternative to the classical parallel fuzzy clustering model, Information Sciences 318 (2015) 28–47.

[64] J. M. Leski, Generalized weighted conditional fuzzy clustering, IEEE Transactions on Fuzzy Systems 11 (6) (2003) 709–715.

[65] J. Yu, Q. Cheng, H. Huang, Analysis of the weighting exponent in the FCM, IEEE Transactions on Systems, Man, and Cybernetics, Part B: Cybernetics 34 (1) (2004) 634–639.

[66] M. Trivedi, J. C. Bezdek, Low-level segmentation of aerial images with fuzzy clustering, IEEE Transactions on Systems, Man, and Cybernetics 16 (4) (1986) 589–598.

[67] N. R. Pal, J. C. Bezdek, On cluster validity for the fuzzy C-means model, IEEE Transactions on Fuzzy Systems 3 (3) (1995) 370–379.

[68] H. Frigui, R. Krishnapuram, A robust algorithm for automatic extraction of an unknown number of clusters from noisy data, Pattern Recognition Letters 17 (12) (1996) 1223–1232.

[69] F. Klawonn, R. Kruse, H. Timm, Fuzzy shell cluster analysis, in: G. della Riccia, H. Lenz, R. Kruse (Eds.), Learning, networks and statistics, Springer, 1997, pp. 105–120.

[70] J. C. Bezdek, A physical interpretation of fuzzy ISODATA, IEEE Transactions on Systems, Man and Cybernetics SMC-6 (5) (1976) 387–389.

[71] F. Klawonn, F. Hoppner, What is fuzzy about fuzzy clustering? Understanding and improving the concept of the fuzzifier, in: M. R. Berthold, H.-J. Lenz, E. Bradley, R. Kruse, C. Borgelt (Eds.), Advances in Intelligent Data Analysis V, Vol. 2810 of Lecture Notes in Computer Science, Springer Berlin Heidelberg, 2003, pp. 254–264.

[72] J. K. Lindsey, Comparison of probability distributions, Journal of the Royal Statistical Society. Series B (Methodological) 36 (1) (1974) 38–47.

[73] G. J. Klinker, S. A. Shafer, T. Kanade, The measurement of highlights in color images, International Journal of Computer Vision 2 (1988) 7–32.

[74] A. Abadpour, S. Kasaei, Color PCA eigenimages and their application to compression and watermarking, IEE Image & Vision Computing 26 (7) (2008) 878–890.